Automotive Technician Training: Entry Level 3

A blended learning approach to automotive engineering at foundation level

Used alongside the ATT Training online learning resources, this textbook covers everything that students need to learn in order to pass Introduction to Motor Vehicle Engineering (EL3) automotive courses. This book takes a blended learning approach, using interactive features that make learning more enjoyable as well as more effective.

When linked with the ATT Training online resources, it provides a comprehensive package that includes activities, animations, assessments and further reading. Information and activities are set out in sequence, to meet teacher and learner needs as well as qualification requirements.

Tom Denton is the leading UK automotive author with a teaching career spanning lecturer to head of automotive engineering in a large college. His range of automotive textbooks published since 1995 are bestsellers and led to his authoring of the Automotive Technician Training multimedia system that is in common use in the UK, USA and several other countries. Tom now works as the eLearning Development Manager for the Institute of the Motor Industry (IMI).

Automotive Technician Training: Entry Level 3

Introduction to Light Vehicle Technology

Tom Denton

LONDON AND NEW YORK

First published 2015
by Routledge
2 Park Square, Milton Park, Abingdon, Oxon OX14 4RN

and by Routledge
711 Third Avenue, New York, NY 10017

Routledge is an imprint of the Taylor & Francis Group, an informa business

© 2015 Tom Denton

The right of Tom Denton to be identified as author of this work has been asserted by him/her in accordance with sections 77 and 78 of the Copyright, Designs and Patents Act 1988.

All rights reserved. No part of this book may be reprinted or reproduced or utilised in any form or by any electronic, mechanical, or other means, now known or hereafter invented, including photocopying and recording, or in any information storage or retrieval system, without permission in writing from the publishers.

Trademark notice: Product or corporate names may be trademarks or registered trademarks, and are used only for identification and explanation without intent to infringe.

British Library Cataloguing-in-Publication Data
A catalogue record for this book is available from the British Library

Library of Congress Cataloging in Publication Data
[CIP data]

ISBN: 978-0-415-72040-3 (pbk)
ISBN: 978-1-315-85889-0 (ebk)

Typeset in Univers LT by
Servis Filmsetting Ltd, Stockport, Cheshire

Contents

Preface ix
Acknowledgement x

1 ATT interactive **1**
1.1 Introduction 2
1.2 Learning activities 4
1.3 ATT interactive puzzles 11
1.3.1 Cryptograms 11
1.3.2 Anagrams 11
1.3.3 Word search 11
1.3.4 Crosswords 12

2 Working safely **13**
2.1 Introduction to motor vehicle workshop safety 13
2.1.1 What you must know about motor vehicle workshop safety 14
2.1.2 Personal protective equipment 15
2.1.3 Moving loads 16
2.1.4 Working environment 18
2.1.5 Equipment maintenance 18
2.1.6 Hazards 21
2.1.7 Fire 24
2.1.8 Signage 26
2.1.9 Safety procedures 28
2.2 Working safely puzzles 29
2.2.1 Cryptograms 29
2.2.2 Anagrams 29
2.2.3 Word search 29
2.2.4 Crosswords 30

3 Automotive industry **31**
3.1 Introduction to the retail automotive maintenance and repair industry 31
3.1.1 What you must know about the retail automotive maintenance and repair industry 32
3.1.2 The motor trade introduction 33
3.1.3 Companies 37
3.2 EL13 Light Vehicle Construction 41
3.2.1 What you must know about vehicle construction 41
3.2.2 Layouts 42
3.2.3 Body design 45

3.3 Automotive industry puzzles 47
3.3.1 Cryptograms 47
3.3.2 Anagrams 47
3.3.3 Word search 47
3.3.4 Crosswords 48

4 Workshop skills **49**
4.1 Introduction to workshop tools and equipment 49
4.1.1 What you must know about workshop tools and equipment: 50
4.1.2 Hand tools 51
4.1.3 Measurement 57
4.1.4 Workshop equipment 65
4.1.5 Nuts, screws, washers and bolts 73
4.2 Workshop skills puzzles 81
4.2.1 Cryptograms 81
4.2.2 Anagrams 81
4.2.3 Word search 82
4.2.4 Crosswords 83

5 Maintenance **85**
5.1 Routine vehicle checks 85
5.1.1 What you must know about vehicle checks 86
5.1.2 Main systems 87
5.1.3 Service sheets 90
5.1.4 Effects of incorrect adjustments 91
5.1.5 Information sources 92
5.2 Routine vehicle maintenance processes and procedures 94
5.2.1 What you must know about vehicle inspection 95
5.2.2 Introduction 96
5.2.3 Maintenance and inspections 98
5.3 Basic vehicle valeting 100
5.3.1 What you must know about valeting 101
5.3.2 Overview, equipment and safety 102
5.3.3 Exterior cleaning 105
5.3.4 Interior cleaning 109
5.4 Maintenance puzzles 110
5.4.1 Cryptograms 110
5.4.2 Anagrams 111
5.4.3 Word search 111
5.4.4 Crosswords 112

6 Engine systems **113**
6.1 Principles of engine components and operation 113
6.1.1 What you must know about engine components and operation 114
6.1.2 Operating cycles 114
6.1.3 Cylinder components 120
6.1.4 Valves and valve gear 122
6.1.5 Engine electrical 125

6.2 Routine cooling and lubrication system checks 127
6.2.1 What you must know about lubrication and cooling 128
6.2.2 Cooling introduction 128
6.2.3 Cooling components 133
6.2.4 Friction and lubrication 137
6.2.5 Oils and specifications 138
6.3 Introduction to spark ignition fuel systems 139
6.3.1 What you must know about spark ignition fuel systems 140
6.3.2 Fuel supply 141
6.3.3 Petrol injection 142
6.4 Introduction to compression ignition fuel systems 154
6.4.1 What you must know about compression ignition fuel systems 155
6.4.2 Diesel injection 156
6.4.3 Exhaust systems 162
6.5 Introduction to Vehicle Ignition Systems 165
6.5.1 What you must know about ignition systems 165
6.5.2 Ignition overview 166
6.5.3 Spark plugs and leads 170
6.6 Engine systems puzzles 173
6.6.1 Cryptograms 173
6.6.2 Anagrams 173
6.6.3 Word search 173
6.6.4 Crosswords 174

7 Electrical systems **175**
7.1 Vehicle lighting system maintenance 175
7.1.1 What you must know about lighting systems 176
7.1.2 Electrical components and circuits 177
7.1.3 Lighting systems 184
7.1.4 Stoplights and reverse lights 190
7.2 Electrical systems puzzles 193
7.2.1 Cryptograms 193
7.2.2 Anagrams 193
7.2.3 Word search 194
7.2.4 Crossword 195

8 Chassis systems **197**
8.1 Principles of light vehicle steering and suspension systems 197
8.1.1 What you must know about steering and suspension: 198
8.1.2 Suspension 199
8.1.3 Check damper operation 212
8.1.4 Steering 212
8.1.5 Steering alignment 219
8.2 Routine braking system checks 222
8.2.1 What you must know about braking systems 223
8.2.1 Brakes introduction 224
8.2.2 Disc, drum and parking brakes 227
8.3 Routine wheel and tyre checks 232
8.3.1 What you must know about wheels and tyres 233

8.3.2 Wheels 234
8.3.3 Tyres 240
8.4 Chassis systems puzzles 245
8.4.1 Cryptograms 245
8.4.2 Anagrams 245
8.4.3 Word search 245
8.4.4 Crosswords 246

9 Transmission systems **247**
9.1 L111 Introduction to Vehicle Transmission Systems 247
9.1.1 What you must know about transmission systems 248
9.1.2 Clutch 249
9.1.3 Gearbox 253
9.1.4 Automatic transmission 259
9.1.5 Propshafts and driveshafts 264
9.1.6 Final drive 269
9.2 Transmission systems puzzles 271
9.2.1 Cryptograms 271
9.2.2 Anagrams 271
9.2.3 Word search 271
9.2.4 Crossword 272

Index 274

Preface

I hope you find this book useful and informative. Comments, suggestions and feedback are always welcome at our website: www.automotivett.com. You will also find lots of other online resources to help with your studies.

Throughout this book you will find questions, interactive activities, diagrams that need labels adding and more – use the eLearning material to complete these tasks. It is a good way to learn, I promise!

Good luck and I hope you find automotive technology as interesting as I still do.

Tom Denton

Acknowledgements

Over the years, many companies and people have helped in the production of my books. I am therefore very grateful to the following who have supplied information and/or permission to reproduce photographs and/or diagrams:

AA Photo Library
AC Delco Inc.
Alpine Audio Systems Ltd
Autologic Data Systems Ltd
BMW UK Ltd
C&K Components Inc.
Citroën UK Ltd
Clarion Car Audio Ltd
Delphi Automotive Systems Inc.
Eberspaecher GmbH.
Fluke Instruments UK Ltd
Ford Motor Company Ltd
Ford Media
General Motors Media
GenRad Ltd
Hella UK Ltd
Honda Cars UK. Ltd
Hyundai UK Ltd
Lucas UK Ltd
Jack Sealey Ltd
Jaguar Cars Ltd
Kavlico Corp.
Honda UK Ltd
LucasVarity Ltd
Mazda Cars UK Ltd
Mercedes Cars UK Ltd
Mitsubishi Cars UK Ltd
NGK Plugs UK Ltd
Nissan Cars UK Ltd
Peugeot UK Ltd
Philips UK Ltd
Pioneer Radio Ltd
Porsche Cars UK Ltd
Robert Bosch GmbH.
Bosch Media
Robert Bosch UK Ltd
Rover Cars Ltd
Saab Cars UK Ltd
Saab Media
Scandmec Ltd
Toyota Cars UK Ltd
Tracker UK Ltd
Unipart Group Ltd
Valeo UK Ltd
VDO Instruments
Volvo Cars Ltd
ZF Servomatic Ltd
Snap-on Tools Inc.
Sofanou (France)
Sun Electric UK Ltd
Thrust SSC Land Speed Team

If I have used any information or mentioned a company name that is not noted here, please accept my apologies and let me know so it can be rectified as soon as possible.

CHAPTER 1

ATT interactive

After successful completion of this section you will be able to show you have achieved these outcomes:

- Understand the various icons and symbols and structure used in this book and online.
- Understand how to use the learning activities and other features.

Automotive Technician Training: Entry Level 3. 978-0-415-72040-3.
© Tom Denton. Published by Taylor & Francis. All rights reserved.

1.1 Introduction

This textbook should be used in conjunction with the ATT multimedia materials; it is not intended to be a standalone resource. But, it is designed to help you learn . . .

. . . and the best way to learn anything new is to interact with it. In other words, be an active learner. Sitting back being passive and expecting your brain to remember stuff doesn't work!

You will find this button (or something similar) on all our multimedia learning screen – click it to see what happens!

As you work through this book you will see this symbol next to some paragraphs:

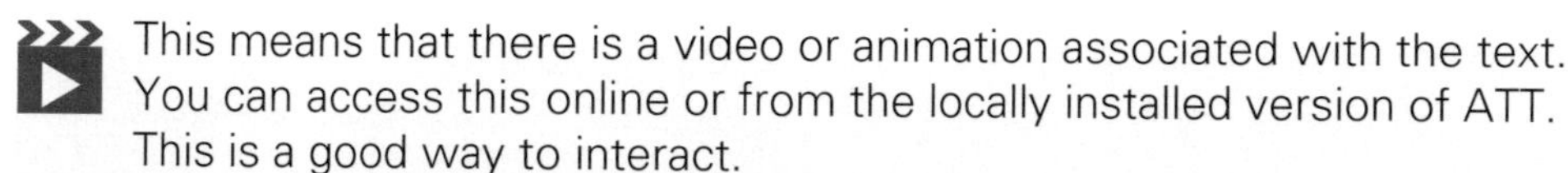

This means that there is a video or animation associated with the text. You can access this online or from the locally installed version of ATT. This is a good way to interact.

The other image you will see at the start of each main chapter, is a QR code (as above). If you have suitable software on your computer (and a camera) or a smartphone app, pointing at this image will link to some useful resources.

Note that the practical sections are not in this book (to save space) but they are all available in full multimedia online.

Online learning link: www.automotivett.org

Screen title Most of the paragraphs of text in this book with start with a title in bold as shown here! This is the title that you will see on the multimedia learning screen. And to help you interact even more and make learning about automotive technology fun as well as interesting, we have developed ATT

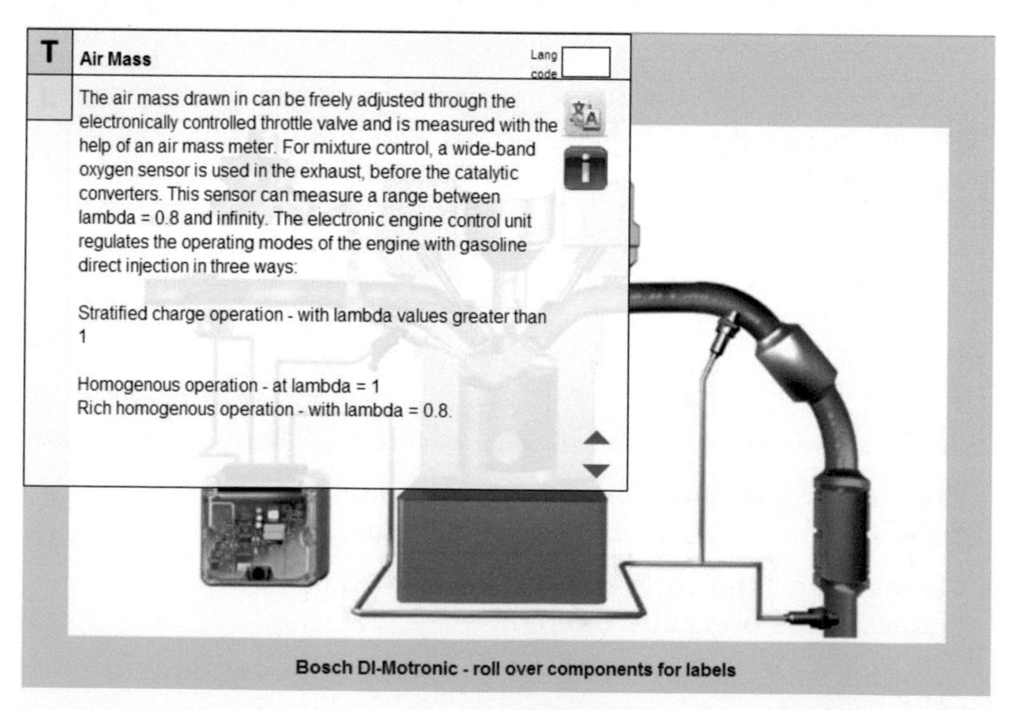

Figure 1.1 Example screen showing the interactive and translate buttons

interactive. All the activities in this book can be carried out with using features on the interactive website. Paper and pencils will work too in most cases too so you can still work if you don't have internet access.

Visit: www.automotivett.org to access a demonstration of the amazing and unique features of ATT *interactive*:

The translate feature available on all our learning screens. Enter the language code in the small box and then click the Google translate button. I used 'es' (Spanish) and the result is shown as Figure 1.2.

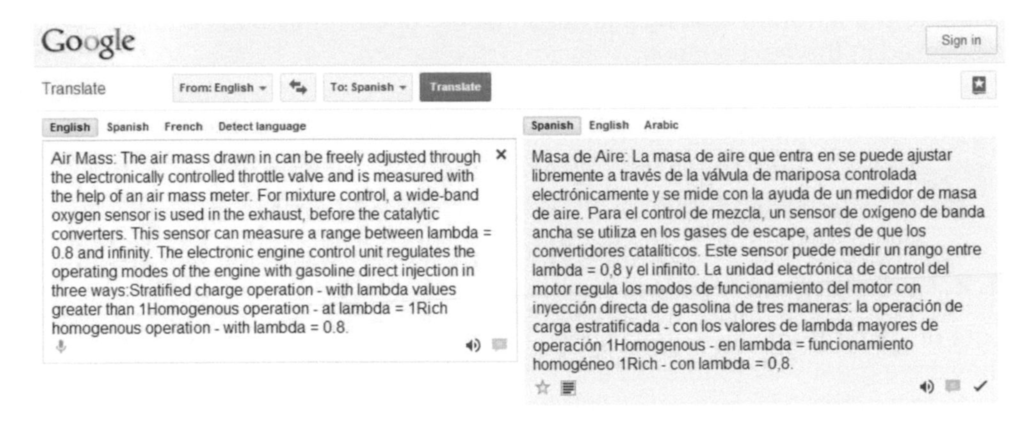

Figure 1.2 Google translate website

Now although this machine translation is very good it is not perfect and so fluent speakers will notice some amusing errors (filing metal and filing papers can be mixed up for example!). However, the feedback we have received is that it is a great help for people working in English as a second language.

Figure 1.3 ATT interactive website

Clicking the blue interactive button takes you to the website shown above. We are working on this all the time so new options may be available but it will be very similar.

You can also translate at this point and if you entered a language code then the search features will use that language. If not it will default to English.

Clicking one of the buttons on the site may take you to an external website where you will appreciate we cannot be responsible for the content! There are currently twelve main headings and some of these have several associated buttons. Experiment as much as you like. These 12 options also relate to the different leaning activities that are used in this book and are discussed in the next section.

Now complete the multiple choice quiz associated with this topic/ subject area.

1.2 Learning activities

Lots of learning activities are included in the ATT textbooks, and remember they are a great way to *interact* and learn. In most cases the best option will be to follow the ATT *interactive* link from a learning screen. The answers to questions or notes/bullet/labels and other activities can be written directly in the textbook, on a separate notebook or perhaps even better as a document or in an electronic notebook.

Some activities will just show an icon (add labels to a diagram for example) others will give further instructions and maybe a space for the work to be completed. For all the activities in this section I have included an example of what you could do.

Tip: Pressing PrtScr will copy the screen to the clipboard for pasting in your electronic notebook (Word document or whatever). Better still, use the annotator supplied as part of the ATT system or available from http:// getgreenshot.org

Note: The activities suggested in the text are recommended, but you can do different types – and, remember, the more you do, the more you will learn!

Information search This activity will usually be something similar to this:

Use a library or the interactive web search tools to examine the subject in this section in more detail.
Here is the link I found useful after searching on the interactive site for one of the key words in the text: http://en.wikipedia.org/wiki/ Catalytic_converters

Looking in other textbooks in a library is a good way to see the subject explained in a different way. Perhaps even better is to user the search options on the interactive site. Clicking any of the buttons in this row (see Figure 1.3) will search for the screen title, 'Air Mass' in this example, or you can select a word or phrase in the text and it will search or that instead.

Media search This activity is very similar to the previous except it asks you to search for images and videos.

Use the interactive media search tools to look for pictures and videos to examine the subject in this section in more detail.
Searching for Catalytic converter I found this rather nice image on Flickr:

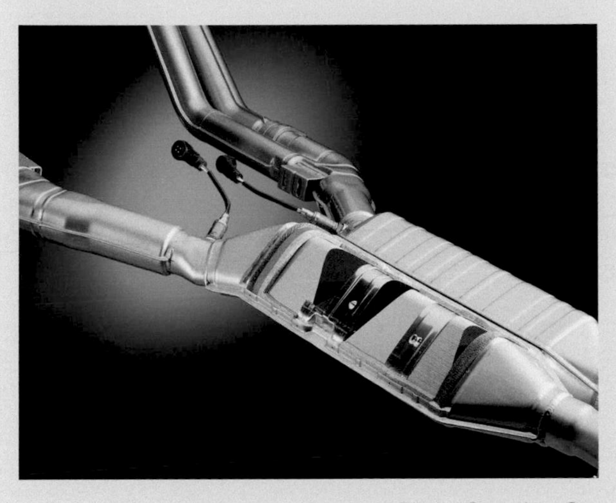

Figure 1.4 Daimler catalytic converter (Source: Flickr Creative Commons)

The buttons in this row search for the screen title or selected text in the same way as before.

Word cloud A word cloud shows the most common words in a block of text in a larger font. It is a great way to focus in on the important aspects of a learning screen or paragraph of text. There are a few different options available on the interactive site. Here is an example of a word cloud I created based on the example screen shown earlier. Note that clicking the interactive button also copies the text onto the clipboard for pasting elsewhere.

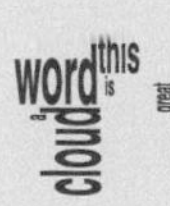

Create a word cloud for one or more of the most important screens or blocks of text in this section.

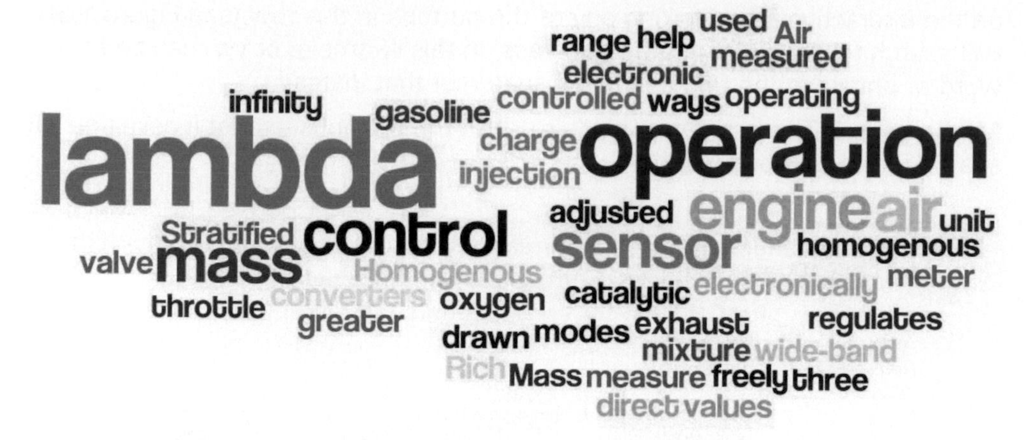

Figure 1.5 Example of a word cloud created by the amazing 'Wordle' website

Word puzzles Crossword and word search puzzles are a great way to learn new important words and the associated technologies. A good method is to work in pairs so you each create a puzzle and then swap and try to complete the answers.

Construct a crossword puzzle using important words from this section. Hint: use the ATT glossary where you can copy the words and definitions (clues!). About 20 words is a good puzzle.
OR
Construct a word search grid using some key words from this section. About 10 words in a 12 × 12 grid is usually enough.

There is also an option for creating anagrams in this row of interactive features – not sure how useful that is but it is fun! Here is a crossword I prepared earlier:

Across

3 To join in with an activity is to be . . .

5 Mode of transport

6 Technical reading material

Down

1 Look for connected letters in a grid

2 What you are doing now

4 Subject of this book

Figure 1.6 Crossword puzzle (Source: TeachersCorner.net)

Mind map/wall These activities work fine either with pen and paper or by using the online features. Figure 1.7 is an example of a simple mind map I created about brakes using a link on the interactive site.

Create a mind map to illustrate the important features of a component or system in this section.

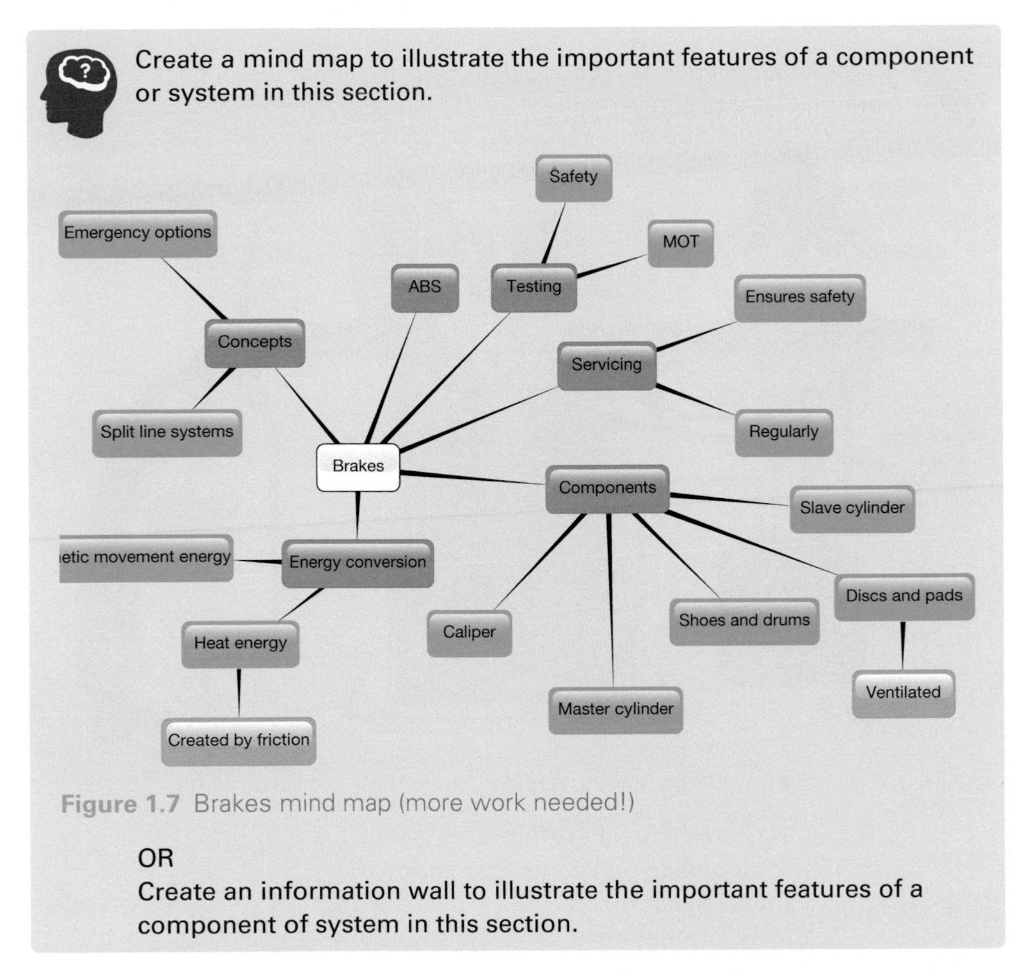

Figure 1.7 Brakes mind map (more work needed!)

OR
Create an information wall to illustrate the important features of a component of system in this section.

Notes/bullets Three great tools for keeping notes electronically are Evernote, Microsoft OneNote and Google drive. My favourite at the moment is OneNote but I find all these tools easy to work with and they can be used online or offline and also synched to or from my smartphone. Of course using any word processor is fine – as is using a pen. The following is an example of some key bullet points relating to an introduction to brakes:

Look back over the previous section and write out a list of the key bullet points.

- *Brakes work by converting movement energy into heat*
- *The foot brake acts on all wheels and the parking brake usually on two*
- *Friction is used to create the heat*
- *Main components are pads and discs or shoes and drums (pads most common)*
- *Hydraulic fluid is used to transfer the pressure from the drivers foot pedal*

Labels Many of the diagrams in this book have numbers but no labels. Use the multimedia version to find out what they are and write them in to the book (or copy the image and do it electronically)

Complete the labels on the diagram by referring to the appropriate ATT multimedia learning screen.

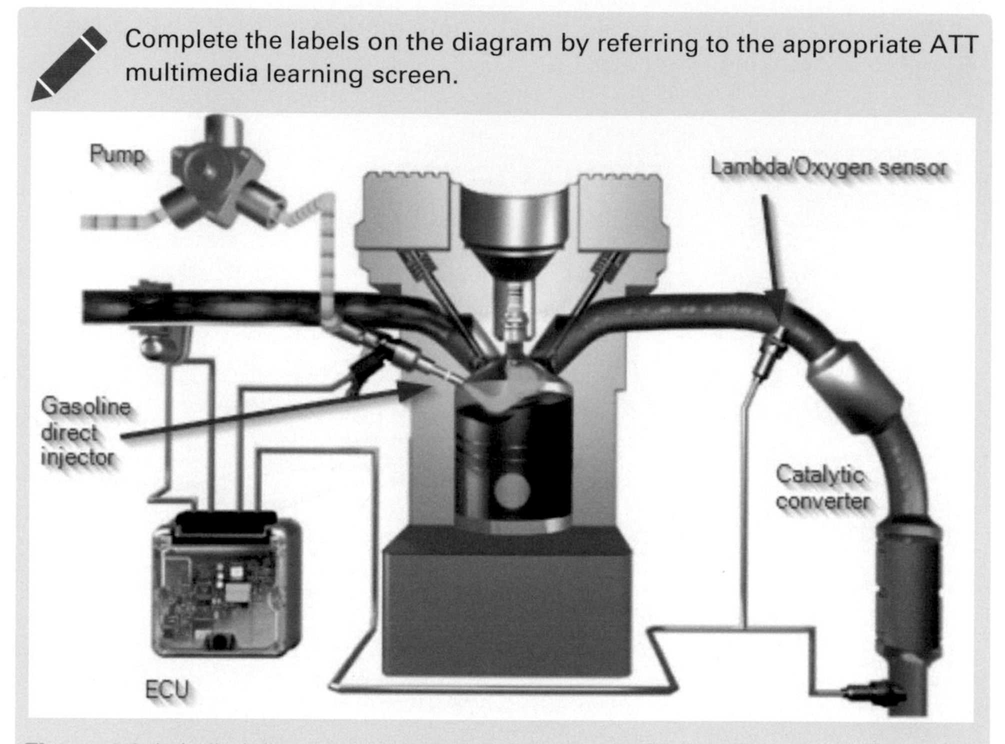

Figure 1.8 Labelled diagrams can be saved in your electronic notebook

OR: List the items separately:

The components on this diagram are:

1 Lambda sensor
2 Catalytic converter
3 High pressure pump
4 Injector

I added labels and the arrows to the diagram above using the Greenshot annotator.

Social If working on a college or school network you may not be allowed to access these sites but you certainly can in your own time. It is a great way to keep in touch, share ideas with your mates and communicate with us here at ATT.

Follow the tweets, Facebook and blog posts from our automotive website linked from www.automotivett.org. You could also set up a Facebook discussion group to talk about specific automotive technology subjects.

Questions Short-answer questions are used at the end of all the technology sections of this book. Write the answers in the box provided or keep them electronically and note the page number so you can refer back to them.

Answer the following questions either here in your book or electronically.

1 What is the address of the ATT web site?
www.automotivett.org
2 What is your favourite type of car?
One that starts every time and is comfortable and reliable (oh, and is ideally a Ferrari)

At the end of each main subject area you could carry out the associated multiple-choice test online or on the DVD/offline version. This box will show as a reminder:

Now complete the multiple choice quiz associated with this topic/subject area.

Sketch Making a simple sketch to help you remember how a component or system works is a good way to learn. You can use a pencil or the online features or any drawing program – even word processors have quite good drawing tools now. The sketch below is my representation of a closed loop control system.

Make a simple sketch to show how one of the main components or systems in the section operates.

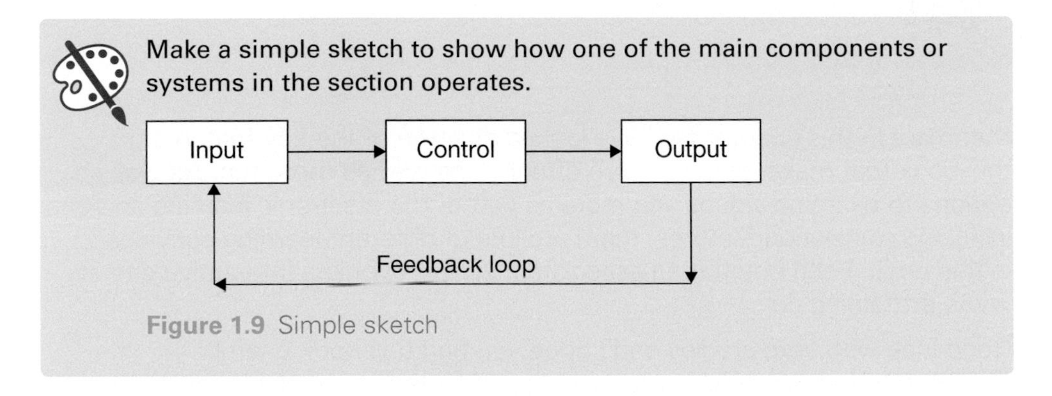

Figure 1.9 Simple sketch

Presentation Preparing and making a presentation to your mates is a great way to learn about something new because you have to study it in detail first! It can be a bit nerve racking at first but is also good fun so don't worry.

There are some great online tools for this or you can use PowerPoint or similar to prepare some slides that you then explain in more detail.

Using images and text, put together a short presentation that you will deliver to your class mates to show how an important component of system from this section works.

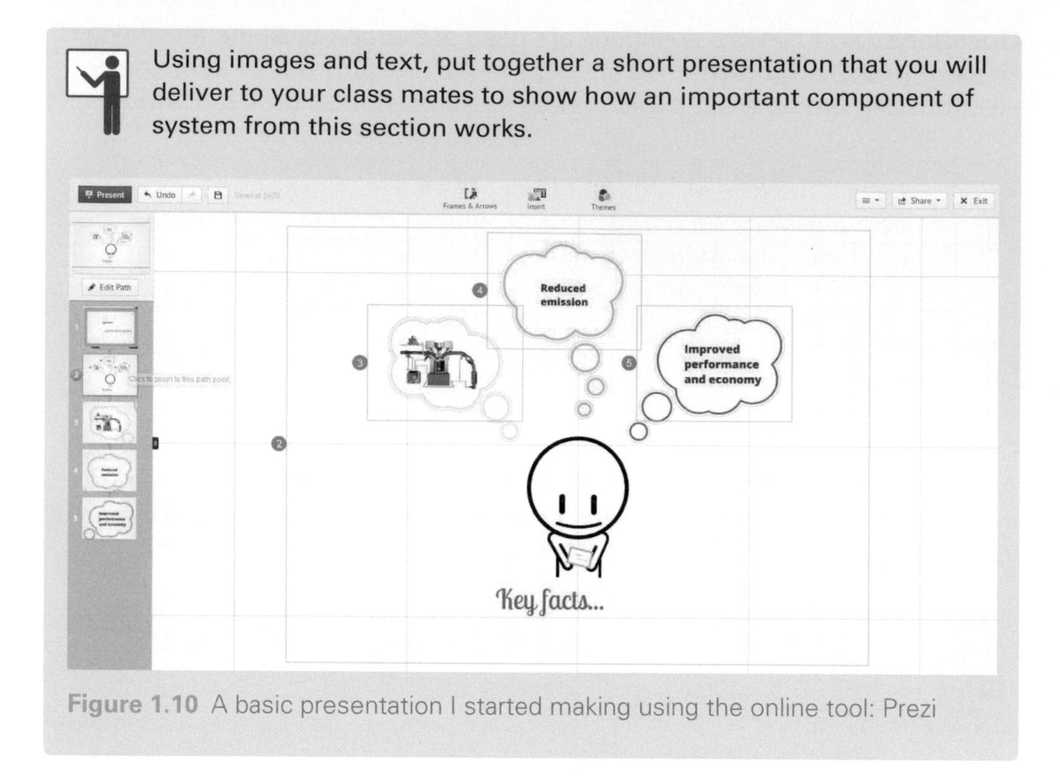

Figure 1.10 A basic presentation I started making using the online tool: Prezi

Practical Clearly practical work is the most important thing we do as automotive technicians.

Refer to the appropriate worksheets and carry out the practical task(s) related to this section – when directed by your teacher or instructor.

Summary In this chapter we have looked at some of the key features of this book that make learning more effective as well as more fun. As well as options to examine videos and more as part of the electronic learning screens, including translation features, there are lots of different learning activities in this book. Each one has an associated link in the online interactive site at www.atttraining.co

Good luck with your studies and I hope you find this book useful. Remember: get involved in your learning and interact – and it is much more interesting!

Now complete the multiple choice quiz associated with this topic/ subject area.

1.3 ATT interactive puzzles

1.3.1 Cryptograms

A	B	C	D	E	F	G	H	I	J	K	L	M	N	O	P	Q	R	S	T	U	V	W	X	Y	Z
	23							11			13														

L						B			I								I		
13	10	25	17	9		23	14		11	9	20	10	17	25	24	20	11	9	5

Cryptogram 1.1 Get involved and know more:

1.3.2 Anagrams

Arched Rows Charred Sow
Crashed Row Scarred Who

Anagrams 1.1 Looking for something in a grid: What word did I use to get these anagrams?

1.3.3 Word search

I T A V E Z R R C Q V G P R E
Y N U O X T L K P Q U Y K I Y
I E T L J U I V T B H Z N E A
Q F O E H F R S N O D W G L I
E A M K R C C O B I B L K X M
W C O J X A S P N E X H T G Y
P W T Y A C C T C C W R E Y X
X D I X S V E T C P N E A L Q
G S V E J R P B I Z Q P Y U B
X T E U N A M X C V Z A D T F
V B S E K E S C L W E P J Y P
M O T N V M Y P E X K K P H H
O V X E I N W L G V B P R B Y
L A Q P R Y G F V S I S F H W
P G B D G E F S L I C N E P E

AUTOMOTIVE
INTERACTIVE
INTERNET
PAPER
PENCILS
WEBSITE

Word search 1.1 Introduction

1.3.4 Crosswords

Across

1 Combination of different resources such as sound, text and images

4 Image that can be scanned to connect to a resource automatically (2, 4)

5 Automotive technician training (1, 1, 1)

8 Finding out something new

9 Convert to another language

10 Pages on the internet

Down

2 Getting involved with something

3 Working with your hands

6 Static picture

7 Moving images

Crossword 1.1 Introduction

CHAPTER 2

Working safely

2.1 Introduction to motor vehicle workshop safety

Together with the multimedia resources, this section is ideal material for students working towards the IMI Awards unit EL01, the safety aspects of City and Guilds entry level units and all other similar foundation or introduction level qualifications.

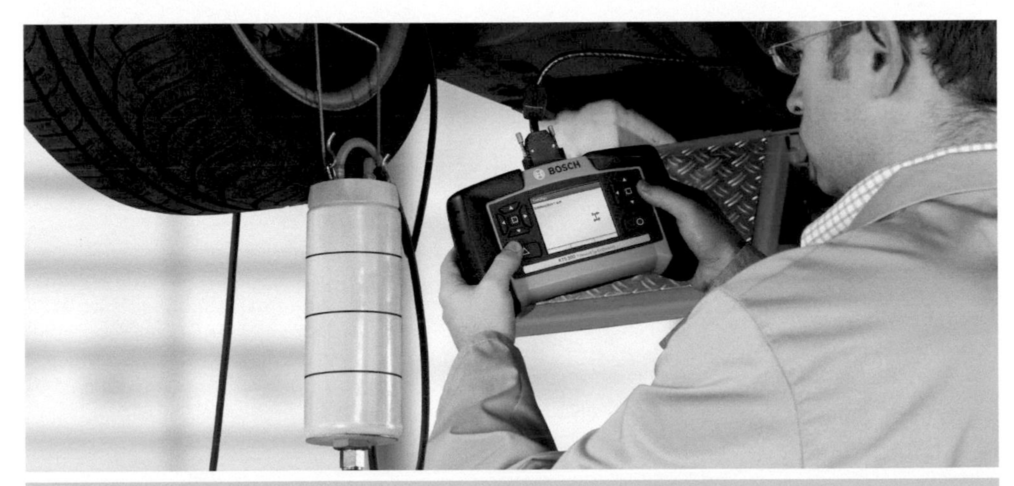

After successful completion of this section you will be able to show you have achieved these outcomes:

- Be able to demonstrate safe working practices within a motor vehicle workshop.
- Know the essential personal protective equipment required while working in a motor vehicle workshop.
- Know a range of potential hazards in a motor vehicle workshop.
- Know what to do in the event of an accident or emergency.
- Be able to demonstrate their awareness of environmental protection issues and safe disposal of hazardous substances.

Automotive Technician Training: Entry Level 3. 978-0-415-72040-3.
© Tom Denton. Published by Taylor & Francis. All rights reserved.

2.1.1 What you must know about motor vehicle workshop safety:

Safe use and storage of the following equipment: jacks, ramps, axle stands, air lines and attachments, electrical apparatus, general hand tools, tracking gauges, tyre changing equipment, pressure testing equipment etc.

Using the correct PPE is important: overalls, gloves, safety boots or shoes, wellingtons, wet working equipment, goggles, face shields and breathing masks

Fire equipment and procedures are very important: types of fire extinguisher, classes of fire, signage, annual maintenance, who may or may not use firefighting equipment, consequences of incorrect fire extinguisher being used and the most common extinguisher used in workshop and the reason why

You must be able to identify and where appropriate dispose of things in an environmentally friendly way: diesel and petrol, oils, paints, cleaning solutions, aerosols, flammable materials (rags, paper etc.), gas and other pressurised cylinders

Common workshop hazards are: tripping, fire, noise, dust, working at height, poor work practices, fooling around, running, smoking, eating and drinking, and drugs (illegal and prescribed)

Know what to do in case of an accident: company procedure, raise alarm, notify senior person, follow instructions as directed, your responsibilities to others and the consequences of not following instructions

You need to know: who is the person in charge of first aid, the contents of first aid box, that it should be visible at all times, how to notify senior personnel of problems and that the accident/incident book should be kept safely and accurately

Get to know who these people are in your workplace: mentor, supervisor, senior personnel, employer and manager

2.1.2 Personal protective equipment

Personal protective equipment (PPE), such as safety clothing, is very important to protect yourself. Some people think it clever or tough not to use protection. They are very sad and will die or be injured long before you! Some things are obvious such as when holding a hot or sharp exhaust you would likely be burned or cut! Other things such as breathing in brake dust, or working in a noisy area, do not produce immediately noticeable effects but could affect you later in life.

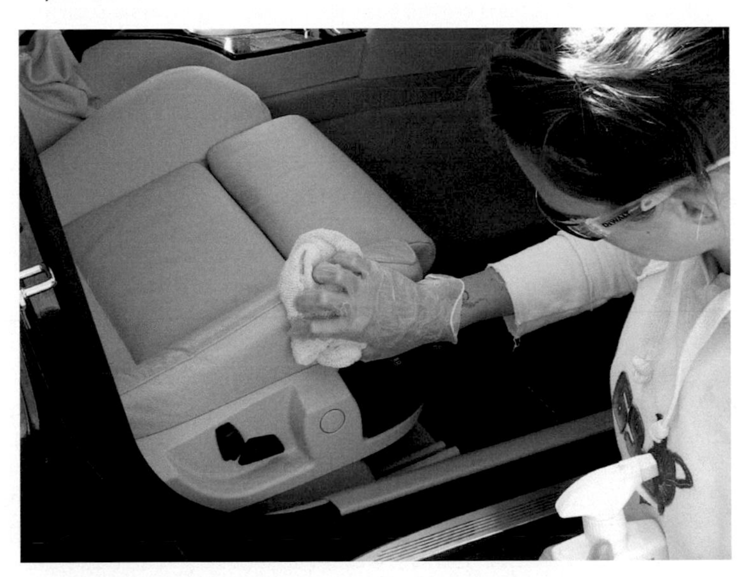

Figure 2.1 Eye protection and gloves in use

Fortunately the risks to workers are now quite well understood and we can protect ourselves before it is too late.

Figure 2.2 Protective clothing for spot welding

Remember! There is a multimedia version of this textbook that includes additional images and interactive features: www.automotivett.org

In the following table, I have listed a number of items classed as PPE together with suggested uses. You will see that the use of most items is plain common sense.

Equipment	Notes	Suggested or examples where used
Ear defenders	Must meet appropriate standards	When working in noisy areas or if using an air chisel
Face mask	For individual personal use only	Dusty conditions. When cleaning brakes or if preparing bodywork
High visibility clothing	Fluorescent colours such as yellow or orange	Working in traffic such as when on a breakdown
Leather apron	Should be replaced if it is holed or worn thin	When welding or working with very hot items
Leather gloves	Should be replaced when they become holed or worn thin	When welding or working with very hot items and also if handling sharp metalwork
Life jacket	Must meet current standards	Use when attending vehicle breakdowns on ferries!
Overalls	Should be kept clean and be flame proof if used for welding.	These should be worn at all times to protect your clothes and skin. If you get too hot just wear shorts and a T-shirt underneath
Rubber or plastic apron	Replace if holed	Use if you do a lot of work with battery acid or with strong solvents
Rubber or plastic gloves	Replace if holed	Gloves must always be used when using degreasing equipment
Safety shoes or boots	Strong toe caps are recommended	Working in any workshop with heavy equipment
Safety goggles	Keep the lenses clean and prevent scratches	Always use goggles when grinding or when any risk of eye contamination. Cheap plastic goggles are much easier to come by than new eyes
Safety helmet	Must be to current standards	Under vehicle work in some cases
Welding goggles or welding mask.	Check the goggles are suitable for the type of welding. Gas welding goggles are NOT good enough when arc welding	You should wear welding goggles or use a mask even if you are only assisting by holding something

2.1.3 Moving loads

Injuries in a workshop are often due to incorrect lifting or moving of heavy loads. In motor vehicle workshops, heavy and large components, like engines and gearboxes can cause injury when being removed and refitted.

A few simple precautions will prevent you from injuring yourself, or others.

- Never try to lift anything beyond your capability – get a mate to help. The amount you can safely lift will vary but any more than you feel comfortable with, you should get help.
- Whenever possible use an engine hoist, a transmission jack or a trolley jack.
- Lift correctly, using the legs and keep your back straight.
- When moving heavy loads on a trolley, get help and position yourself so you will not be run over if you lose control.

Figure 2.3 Heavy load – correct lifting method

The ideal option in all cases is simply to avoid manual handling where possible.

Figure 2.4 Engine crane (Source: Blue-Point)

Now complete the multiple choice quiz associated with this topic/subject area.

2.1.4 Working environment

There are three main reasons for keeping your workshop and equipment clean and tidy:

1 It makes it a safer place to work.
2 It makes it a better place to work.
3 It gives a better image to your customers.

Servicing and fixing motor vehicles is in some cases a dirty job. But if you clean up after any dirty job then you will find your workshop a much more pleasant place to work.

- The workshop and floor should be uncluttered and clean to prevent accidents and fires as well as maintaining the general appearance.
- Your workspace reflects your ability as a technician. A tidy workspace equals a tidy mind equals a tidy job equals a tidy wage when you are qualified.
- Hand tools should be kept clean as you are working. You will pay a lot of money for your tools, look after them and they will look after you in the long term.
- Large equipment should only be cleaned by a trained person or a person under supervision. Obvious precautions are to ensure equipment cannot be operated while you are working on it and only use appropriate cleaning methods. For example would you use a bucket of water or a brush to clean down an electric pillar drill? I hope you answered 'the brush'!

In motor vehicle workshops many different cleaning operations are carried out. This means a number of different materials are required. It is not possible to mention every brand name here so I have split the materials into three different types. It is important to note that the manufacturer's instructions printed on the container must be followed at all times.

Material	Purpose	Notes
Detergent	Mixed with water for washing vehicles, etc. Also used in steam cleaners for engine washing, etc.	Some industrial detergents are very strong and should not be allowed in contact with your skin.
Solvents	To wash away and dissolve grease and oil, etc. The best example is the liquid in the degreaser or parts washer which all workshops will have.	NEVER use solvents such as thinners or fuel because they are highly inflammable. Suitable PPE should be used, for example gloves, etc. They may attack your skin. Many are flammable. The vapour given off can be dangerous. Serious problems if splashed in to eyes. Read the label.
Absorbent granules	To mop up oil and other types of spills. They soak up the spillage after a short time and can then be swept up.	Most granules are a chalk or clay-type material which has been dried out.

2.1.5 Equipment maintenance

The cleaning and maintenance of equipment plays a big part in good housekeeping. This includes large equipment such as ramps, hoists, etc. to small hand tools. Always remember that no one should clean, maintain or use

Figure 2.5 Hand tools

large equipment unless they have had sufficient training or are working under the supervision of an experienced and qualified person.

- Hand tools are expensive so do look after them and in the long term they will look after you.
- Technicians need to learn and be aware of the following points regarding equipment:
  - select and use equipment for basic hand tool maintenance activities
  - storing hand tools safely and accessibly
  - how to report faulty or damaged work tools and equipment
  - safety when cleaning and maintaining work tools and equipment.

It is important to store hand tools safely. Any hand tool left lying around can be a potential hazard to the unsuspecting person or could cause damage to a customer's vehicle. Always make sure that hand tools are stored correctly in either a tool box or in the designated place. If you think that you are likely to need a particular tool in due cause and don't want to put it back then be aware of where you place it. Obviously you will want it handy but at the same time you need to think of safety.

Safety also applies to the tool you are using. Don't put it down in a place where it can be damaged. Wherever you store or place a hand tool think of the following points:

- Safety of yourself and others.
- Protection of the customer's vehicle.
- Protection of other tools and workshop equipment.
- Protection of the tool itself.

From time to time tools and equipment will develop faults or get damaged, however careful you are with them. If you find any damage to equipment it is your duty as a technician to report it or see if it has already been reported.

Don't leave it to someone else or assume that it must have been spotted by one of your colleagues. It is quite likely that it hasn't.

Figure 2.6 Taking care

Create a mind map to illustrate the important features of a component or system in this section.

Does your workshop have a procedure for reporting faulty equipment? Does it need to be written on a report form for instance? Who is the appropriate person to report this fault to? Your supervisor or perhaps another member of staff. Whoever and however it needs to be done quickly. Delay could possibly make the fault worse but more importantly if it needs to be used, work will be held up. Would this fault be a potential safety issue? Obviously a very important point to consider. Report it immediately so the problem can be fixed.

In the previous section I mentioned the importance of working safely when cleaning or maintaining equipment. It is important to remember that you must never clean or maintain equipment without adequate training or supervision from a qualified and competent person. Even if you are asked politely say 'no' and explain why. It would be quite likely that the person who asked you to do the task is unaware that you do not have the relevant experience.

Look back over the previous section and write out a list of the key bullet points.

Now complete the multiple choice quiz associated with this topic/ subject area.

2.1.6 Hazards

Working in a motor vehicle workshop is a dangerous occupation – if you do not take care. The most important thing is to be aware of the hazards and then it is easy to avoid the danger. The hazards in a workshop are from two particular sources:

From you, such as caused by:

- Carelessness – particularly while moving vehicles
- Drinking or taking drugs – these badly affect your ability to react to dangerous situations
- Tiredness or sickness – these will affect your abilities to think and work safely
- Messing about – most accidents are caused by people fooling about
- Not using safety equipment – you have a duty to yourself and others to use safety equipment
- Inexperience – or lack of supervision. **If in doubt: ask.**

The surroundings in which you work may have:

- bad ventilation
- poor lighting
- noise
- dangerous substances stored incorrectly
- broken or worn tools and equipment
- faulty machinery
- slippery floors
- untidy benches and floors
- unguarded machinery
- unguarded pits.

The following table lists some of the hazards you will come across in a vehicle workshop. Also listed are some associated risks, together with ways we can reduce them. This is called risk management.

Hazard	Risk	Action
Power tools	Damage to the vehicle or personal injury	Understand how to use the equipment and wear suitable protective clothing, for example gloves and goggles
Working under a car on the ramp	1 The vehicle could roll or be driven off the end 2 You can bang your head on hard or sharp objects when working under the car	1 Ensure you use wheel chocks 2 Set the ramp at the best working height, wear protection if appropriate
Working under a car on a jack	The vehicle could fall on top of you	The correct axle stands should be used and positioned in a secure place
Compressed air	Damage to sensitive organs such as ears or eyes. Death, if air is forced through the skin into your blood stream	Do not fool around with compressed air. A safety nozzle prevents excessive air forces

Hazard	Risk	Action
Dirty hands and skin	Oil, fuel and other contaminants can cause serious health problems. This can range from dermatitis to skin cancer	Use gloves or a good-quality barrier cream and wash your hands regularly. Do not allow dirt to transfer to other parts of your body. Good overalls should be worn at all times
Exhaust fumes	Poisonous gases such as carbon monoxide can kill. The other gases can cause cancer or at best restrict breathing and cause sore throats	Only allow running engines in very well ventilated areas or use an exhaust extraction system
Engine crane	Injury or damage can be caused if the engine swings and falls off	Ensure the crane is strong enough (do not exceed its safe working load SWL). Secure the engine with good-quality sling straps and keep the engine near to the floor when moving across the workshop
Cleaning brakes	Brake dust (especially older types made of asbestos) is dangerous to health	Only wash clean with proper brake cleaner
Fuel	Fire or explosion	Keep all fuels away from sources of ignition. Do not smoke when working on a vehicle
Degreaser solvent	Damage to skin or damage to sensitive components	Wear proper gloves and make sure the solvent will not affect the items you are washing
Spillage such as oil	Easy to slip over or fall and be injured	Clean up spills as they happen and use absorbent granules
Battery electrolyte (acid)	Dangerous on your skin and in particular your eyes. It will also rot your clothes	Wear protective clothing and take extreme care
Welding a vehicle	The obvious risks are burns, fire and heat damage but electric welders such as a MIG welder, can damage sensitive electronic systems	Have fire extinguishers handy, remove combustible materials such as carpets and ensure fuel pipes are nowhere near. The battery earth lead must also be disconnected. Wear gloves and suitable protective clothing such as a leather jacket
Electric hand tools	The same risk as power tools but also the danger of electric shocks, particularly in damp or wet conditions. This can be fatal	Do not use electric tools when damp or wet. Electrical equipment should be inspected regularly by a competent person
Driving over a pit	Driving into the pit	The pit should be covered or have another person help guide you and drive very slowly
Broken tools	Personal injury or damage to the car. For example a file without a handle can stab into your wrist. A faulty ratchet could slip	All tools should be kept in good order at all times. This will also make the work easier
Cleaning fluids	Skin damage or eye damage	Wear gloves and eye protection and also be aware of exactly what precautions are needed by referring to the safety data Information

SAFETY DATA SHEET

ENGINE AND MACHINE CLEANER

Page 1

Issued: 10/07/09

Revision No: 7

1. IDENTIFICATION OF THE SUBSTANCE / PREPARATION AND OF THE COMPANY / UNDERTAKING

Product name: ENGINE AND MACHINE CLEANER

Product code: EMC/1261/040609

Company name: AUTOGLYM
Works Road
Letchworth
Herts
SG6 1LU
UK
Tel: +44 (0) 1462 677766
Fax: +44 (0) 1462 677712
Emergency tel: +44 (0) 1462 489498
Email: sds@autoglym.co.uk

2. HAZARDS IDENTIFICATION

Main hazards: No significant hazard.

3. COMPOSITION / INFORMATION ON INGREDIENTS

Hazardous ingredients: DISODIUM METASILICATE 1-10%
EINECS: 229-912-9 CAS: 6834-92-0
[C] R34; [Xi] R37

- GLYCERINE 1-10%
EINECS: 200-289-5 CAS: 56-81-5
- C9-11 ALCOHOL ETHOXYLATE 1-10%
CAS: 68439-45-2
[Xn] R22; [Xi] R41
- TETRASODIUM N,N-BIS(CARBOXYLATOMELHYL)-L-GLUTAMATE 1-10%
EINECS: 257-573-7 CAS: 51981-21-6
[Xi] R36/38

4. FIRST AID MEASURES (SYMPTOMS)

Skin contact: There may be mild irritation at the site of contact.

Eye contact: There may be irritation and redness.

Ingestion: There may be irritation of the throat.

Inhalation: No symptoms.

4. FIRST AID MEASURES (ACTION)

Skin contact: Wash immediately with plenty of soap and water.

Eye contact: Bathe the eye with running water for 15 minutes.

Figure 2.7 Example of a safety data sheet (Source: AutoGlym)

Look back over the previous section and write out a list of the key bullet points.

2.1.7 Fire

Accidents involving fire are very serious. As well as you or a workmate calling the fire brigade (do not assume it has been done), three simple rules will help you know what to do:

1. Get safe yourself, contact the emergency services – and shout FIRE!
2. Help others to get safe if it does not put you or others at risk.
3. Fight the fire if it does not put you or others at risk.

Of course far better than the above situation is to not let a fire start in the first place.

The fire triangle or combustion triangle is a simple model for understanding the ingredients necessary for most fires. The triangle illustrates a fire requires three elements: heat, fuel and an oxidising agent (usually oxygen from the air). The fire is prevented or extinguished by removing any one of them. A fire naturally occurs when the elements are combined in the right mixture.

Figure 2.8 Fire triangle (Source: Wikimedia)

Without sufficient heat, a fire cannot start or continue. Heat can be removed by the application of a substance which reduces the amount of heat available to the fire reaction. This is often water, which requires heat to change from water to steam. Introducing sufficient quantities and types of powder or gas in the flame also reduces the amount of heat available for the fire reaction. Turning off the electricity in an electrical fire removes the ignition source.

Without fuel, a fire will stop. Fuel can be removed naturally, as where the fire has consumed all the burnable fuel, or manually, by mechanically or chemically removing the fuel from the fire. The fire goes out because a lower concentration of fuel vapour in the flame leads to a decrease in energy release and a lower temperature. Removing the fuel therefore decreases the heat.

Without enough oxygen, a fire cannot start or continue. With a decreased oxygen concentration, the combustion process slows. In most cases, there is plenty of air left when the fire goes out so this is commonly not a major factor.

If a fire does happen your workplace should have a set procedure so for example you will know:

- How the alarm is raised.
- What the alarm sounds like.
- What to do when you hear the alarm.
- Your escape route from the building

- Where to go to assemble.
- Who is responsible for calling the fire brigade

There are a number of different types of fire as shown in the following table:

European/ Australian/ Asian	American	Fuel/heat source
Class A	Class A	Ordinary combustibles
Class B	Class B	Flammable liquids
Class C		Flammable gases
Class D	Class D	Combustible metals
Class E	Class C	Electrical equipment
Class F	Class K	Cooking oil or fat

Use the interactive media search tools to look for pictures and videos to examine the subject in this section in more detail.

If it is safe to do so you should try and put out a small fire. Extinguishers and a fire blanket should be provided. Remember, if you remove one side of the fire triangle, the fire will go out. If you put enough water on a fire it will cool down and go out. However spraying water on an electrical circuit could kill you! Spraying water on a petroleum fire could spread it about and make the problem far worse. This means that a number of different fire extinguishers are needed. Internationally there are several accepted classification methods for hand-held fire extinguishers. Each classification is useful in fighting fires with a particular group of fuel.

Fire extinguishers in the UK, and throughout Europe, are red but with a band or circle of a second colour covering between 5–10% of the surface area of the extinguisher to indicate its contents. Prior to 1997, the entire body of the fire extinguisher was colour coded.

Type	Old code		BS EN 3 colour code	Suitable for use on fire classes (brackets denote sometimes applicable)					
Water	Signal red		Signal red	A					
Foam	Cream		Red with a cream panel above the operating instructions	A	B				
Dry powder	French blue		Red with a blue panel above the operating instructions	(A)	B	C		E	
Carbon dioxide CO2	Black		Red with a black panel above the operating instructions		B			E	
Wet chemical	Not yet in use		Red with a canary yellow panel above the operating instructions	A	(B)				F
Class D powder	French blue		Red with a blue panel above the operating instructions				D		
Halon 1211/ BCF	Emerald Green		No longer in general use	A	B			E	

Figure 2.9 CO_2 and water extinguishers and information posters

In the UK the use of Halon gas is now prohibited except under certain situations such as on aircraft and by the military and police.

2.1.8 Signage

A key safety aspect is to first identify hazards, and then remove them or, if this is not possible, reduce the risk as much as possible and bring the hazard to everyone's attention. This is usually done by using signs or markings. Signs used to mark hazards are often as shown in the following table.

Function	Example	Back colour	Fore colour	Sign
Hazard warning	Danger of electric shock	Yellow	Black	Figure 2.10 Electricity
Mandatory	Use ear defenders when operating this machine	Blue	White	Figure 2.11 Wear ear protection

Function	Example	Back colour	Fore colour	Sign
Prohibition	Not drinking water	White	Red/black	**Figure 2.12** Not drinking water
First aid (escape routes are a similar design)	Location of safety equipment such as first aid	Green	White	**Figure 2.13** First aid
Fire	Location of fire extinguishers	Red	White	**Figure 2.14** Extinguisher
Recycling	Recycling point	White	Green	**Figure 2.15** The three Rs of the environment

Construct a word search grid using some key words from this section. About 10 words in a 12x12 grid is usually enough.

Now complete the multiple choice quiz associated with this topic/subject area.

Remember! There is a multimedia version of this textbook that includes additional images and interactive features: www.automotivett.org

2.1.9 Safety procedures

When you know the set procedures to be followed, it is easier to look after yourself, your workshop and your workmates. You should know:

- Who does what during an emergency.
- The fire procedure for your workplace.
- About different types of fire extinguisher and their uses.
- The procedure for reporting an accident.

If an accident does occur in your workplace the first bit of advice is: keep calm and don't panic! The HASAW states that for companies above a certain size:

- First aid equipment must be available.
- Employers should display simple first aid instructions.
- Fully trained first aiders must be employed.

In your own workplace you should know about the above three points. The following table a guide as to how to react if you come across a serious accident.

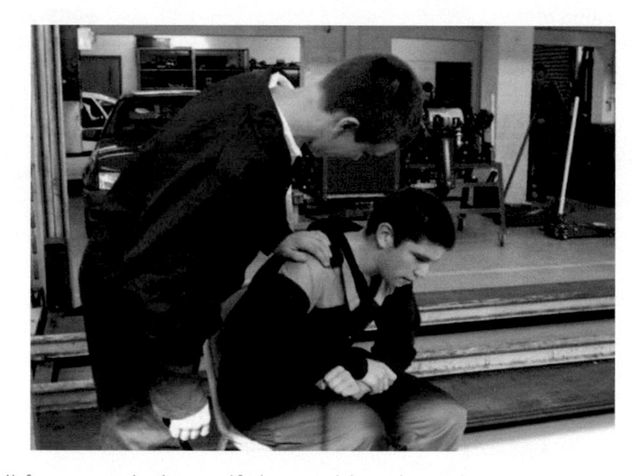

Figure 2.16 Call for an ambulance if the accident is serious

Action	Notes
Assess the situation	Stay calm, a few seconds to think is important
Remove the danger	If the person was working with a machine, turn it off. If someone is electrocuted, switch off the power before you hurt yourself. Even if you are unable to help with the injury you can stop it getting worse
Get help	If you are not trained in first aid, get someone who is and/or phone for an ambulance
Stay with the casualty	If you can do nothing else, the casualty can be helped if you stay with him. Also say that help is on its way and be ready to assist. You may need to guide the ambulance
Report the accident	All accidents must be reported, your company should have an accident book by law. This is a record so that steps can be taken to prevent the accident happening again. Also if the injured person claims compensation, underhanded companies could deny the accident happened
Learn first aid	If you are in a very small company why not get trained now, before the accident?

Look back over the previous section and write out a list of the key bullet points.

2

2.2.0 Working safely puzzles

2.2.1 Cryptograms

A	B	C	D	E	F	G	H	I	J	K	L	M	N	O	P	Q	R	S	T	U	V	W	X	Y	Z
							21	13													24				

H	I		H		V	I		I		I		I									H	I		
21	13	17	21		24	13	6	13	8	13	15	13	26	16		3	15	23	26	21	13	20	17	

Cryptogram 2.1 Be seen:

2.2.2 Anagrams

Caresses Lens Clears Senses
Scales Sneers Scare Lessens

Anagrams 2.1 Be careful, this causes accidents: what word did I use to get these anagrams?

2.2.3 Word search

P	S	E	L	G	G	O	G	G	K	V	N	Z	O	Y	
P	R	E	C	A	U	T	I	O	N	S	P	V	O	T	
M	V	O	D	B	C	X	I	O	E	I	E	T	Z	E	
K	Q	O	T	K	Y	B	P	V	U	R	T	M	J	F	
E	C	F	P	E	F	E	O	G	A	Z	U	F	M	A	
Q	Q	B	F	M	C	L	Z	L	M	U	N	V	I	S	APRON
U	M	G	V	R	G	T	L	A	P	R	O	N	W	L	EQUIPMENT
I	W	A	E	L	L	S	I	S	W	U	L	X	Z	O	GLOVES
P	C	D	S	T	A	N	X	O	B	R	W	X	Z	K	GOGGLES
M	N	W	S	K	R	S	K	R	N	F	D	H	Y	V	LIFTING
E	D	V	X	M	T	P	D	O	N	K	L	O	Y	M	MASK
N	M	T	U	A	D	L	Z	R	O	F	R	S	I	N	OVERALLS
T	J	E	S	L	N	E	M	X	H	X	Q	V	N	G	PRECAUTIONS
Q	X	C	F	V	H	J	I	Y	T	Y	I	D	V	R	PROTECTION
B	C	T	B	V	H	X	P	R	V	D	N	G	G	Z	SAFETY

Word search 2.1 Personal and vehicle protection

2.2.4 Crosswords

Across

1 These should be worn at all times to protect your clothes and skin

3 Breathing in.

9 Regulations to ensure vehicles are built to set standards (12,3,3)

Down

2 Crane used for removing heavy power unit

4 Use these to protect your hearing (3,9)

5 Department of Transport annual safety inspection (1,1,1)

6 Used for our own protection (1,1,1)

7 Eye protection

8 Used to help prevent skin damage when working, for example, with oils (7,6)

10 How your back should be when lifting heavy loads

Crossword 2.1 Personal and vehicle protection

CHAPTER 3

Automobile Industry

3.1 Introduction to the retail automotive maintenance and repair industry

Together with the multimedia resources, this section is ideal material for students working towards the IMI Awards unit EL02, the industrial aspects of City and Guilds entry level units and all other similar foundation or introduction level qualifications.

After successful completion of this section you will be able to show you have achieved these outcomes:

- Know the type of organisations that make up the retail automotive maintenance and repair industry.
- Know the types of vehicle within this sector.
- Know the technical and non-technical job roles available within the sector.

Automotive Technician Training: Entry Level 3. 978-0-415-72040-3.
© Tom Denton Published by Taylor & Francis. All rights reserved.

3.1.1 What you must know about the retail automotive maintenance and repair industry

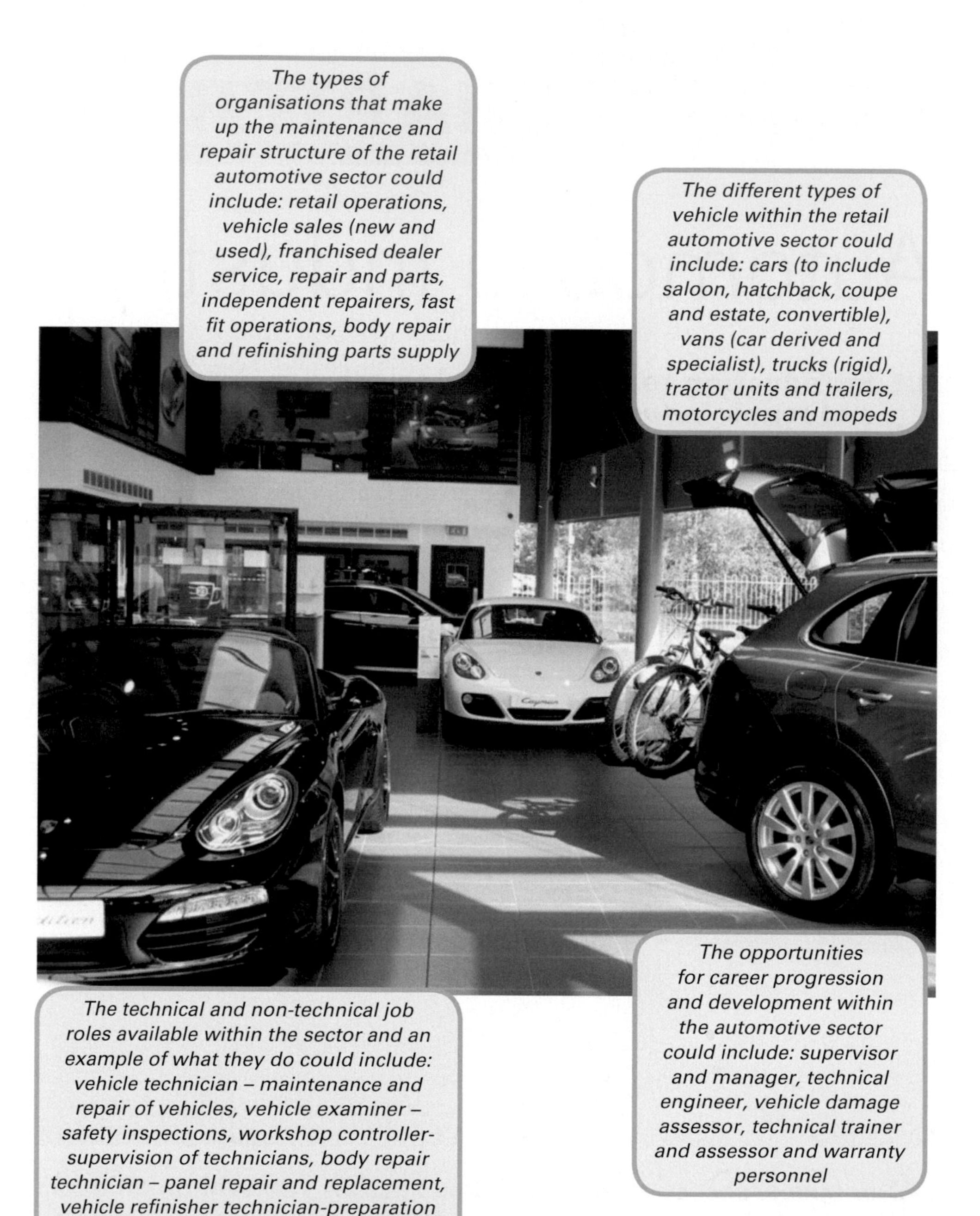

The types of organisations that make up the maintenance and repair structure of the retail automotive sector could include: retail operations, vehicle sales (new and used), franchised dealer service, repair and parts, independent repairers, fast fit operations, body repair and refinishing parts supply

The different types of vehicle within the retail automotive sector could include: cars (to include saloon, hatchback, coupe and estate, convertible), vans (car derived and specialist), trucks (rigid), tractor units and trailers, motorcycles and mopeds

The technical and non-technical job roles available within the sector and an example of what they do could include: vehicle technician – maintenance and repair of vehicles, vehicle examiner – safety inspections, workshop controller-supervision of technicians, body repair technician – panel repair and replacement, vehicle refinisher technician-preparation and painting, parts advisor-supply of parts to public and trade, vehicle sales person – new and used vehicle sales, customer service advisor – liaises with customers and workshop staff, valetor – preparation and cleaning of vehicles, auto-glazing technician – glass repair and replacement

The opportunities for career progression and development within the automotive sector could include: supervisor and manager, technical engineer, vehicle damage assessor, technical trainer and assessor and warranty personnel

3.1.2 The motor trade introduction

Introduction This section will outline some of the jobs that are open to you in the motor trade and help you understand more about the different types of business and how they operate. It is easy to think that the operation of a business does not matter to you. However, I would strongly suggest we should all be interested in the whole business in which we are working. This does not mean to interfere in areas we do not understand. It means we should understand that all parts of the business are important. For example, when you complete a job, enter all the parts used so the person who writes the invoice knows what to charge!

Figure 3.1 **A Volkswagen main dealer**

Figure 3.2 **A Ferrari and Maserati main dealer**

Opportunities The motor trade offers lots of opportunities for those who are willing to work hard and move forwards. There are many different types of job and you will find one to suit you with a little patience and study.

Ask your boss to give you a 'tour' of the garage so that you can appreciate the different tasks carried out and systems that are in place – in particular, make sure you get a reasonable idea about the words and phrases in the table below. If you do not yet have a job you may be able to arrange a visit.

Figure 3.3 **Modern workshop**

Figure 3.4 **Porsche showroom**

Figure 3.5 **Bodyshop**

Key words and phrases Some important words and phrases are presented below.

Customer	The individuals or companies that spend their money at your place of work. This is where your wages come from
Job card	A printed document for recording amongst other things, work required, work done, parts used and the time taken
Invoice	A description of the parts and services supplied with a demand for payment from the customer
Company system	A set way in which things work in one particular company. Most motor vehicle company systems will follow similar rules but will all be a little different
Contract	An offer which is accepted and payment is agreed. If I offer to change your engine oil for £15 and you decide it is a good offer and accept it, we have a contract. This is then binding on us both
Image	This is the impression given by the company to existing and potential customers. Not all companies will want to project the same image
Warranty	An intention that if within an agreed time a problem occurs with the supplied goods or service, it will be rectified free of charge by the supplier
Recording system	An agreed system within a company so that all details of what is requested and/or carried out are recorded. The job card is one of the main parts of this system
Approved repairer	This can mean two things normally. The first is where a particular garage or bodyshop is used by an insurance company to carry out accident repair work. In some cases, however, general repair shops may be approved to carry out warranty work
After sales	A term that applies to all aspects of a main dealer that are involved with looking after a customer's car, after it has been sold to them by the sales team. The service/repair workshop is the best example

Types of MV companies Motor vehicle companies can range from the very small one person businesses to very large main dealers. The systems used by each will be different but the requirements are the same.

Figure 3.6 **Accident repair reception**

Figure 3.7 **Motorist discount store**

Figure 3.8 **Tyre shop**

Figure 3.9 **Ford main dealer**

Figure 3.10 **Good signs are important**

Figure 3.11 **A fast fit centre**

Systems A system should be in place to ensure the level of service provided by the company meets the needs of the customer. The list presented here shows how diverse our trade is.

Mobile mechanics	Servicing and repairs at the owners' home or business. Usually a one-person company.
Bodywork repairers and painters	Specialists in body repair and paintwork
Valeter	These companies specialise in valeting – which should be thought of as much more involved than getting the car washed. Specialist equipment and products are used and proper training is essential.
Fuel stations	These may be owned by an oil company or be independent. Some also do vehicle repair work
Specialised repairers	Auto-electrical, air conditioning, automatic transmission and ICE systems are just some examples.
General repair workshop or independent repairer	Servicing and repairs of most types of vehicles not linked to a specific manufacturer. Often this will be a small business maybe employing two or three people. However, there are some very large independent repairers.
Parts supply	Many companies now supply a wide range of parts. Many will deliver to your workshop.
Fast-fit	Supplying and fitting of exhausts, tyres, radiators, batteries, clutches, brakes and windscreens.
Fleet operator (with workshop)	Many large operators, such as rental companies, will operate their own workshops. Also a large company that has lots of cars, used by sales reps for example, may also have their own workshop and technicians.
Non-franchised dealer	Main activity is the servicing and repairs of a wide range of vehicles, with some sales.
Main dealers or franchised dealers	Usually franchised to one manufacturer, these companies hold a stock of vehicles and parts. The main dealer will be able to carry out all repairs to their own type of vehicle as they hold all of the parts and special tools. They also have access to the latest information specific to their franchise (Ford or Citroën for example). A 'franchise' means that the company has had to pay to become associated with a particular manufacturer but is then guaranteed a certain amount of work and that there will be no other similar dealers within a certain distance.
Multi-franchised dealer	This type of dealer is just like the one above – except they hold more than one franchise – Volvo and GM for example.
Breakdown services	The best known breakdown services are operated by the AA and the RAC. Others include Green Flag and of course many independent garages also offer these roadside repair and recovery services.

Motorists shops	Often described as motorist discount centres or similar, these companies provide parts and materials to amateurs but in some cases also to the smaller independent repairers.

3.1.3 Companies

Company structure A larger motor vehicle company will probably be made up of at least the following departments:

- reception
- workshop
- bodyshop and paint shop
- parts department
- MOT bay
- valeting
- new and second user car sales
- office support
- management
- cleaning and general duties.

Each area will employ one or a number of people. If you work in a very small garage you may have to be all of these people at once! In a large garage it is important that these different areas communicate with each other to ensure that a good service is provided to the customer. The main departments are explained further in the following sections.

Figure 3.12 **Honda reception area**

Figure 3.13 **Different departments on a main dealer site**

Role of a franchised dealer The role of a franchised dealer is to supply local:

- new and used franchised vehicles
- franchise parts and accessories
- repair and servicing facilities for franchise vehicles.

The dealer is also a source of communication and liaison with the vehicle manufacturer.

Reception and booking systems The reception whether in a large or small company is often the point of first contact with new customers. It is very

Figure 3.14 **One of the ATT company cars (we wish!)**

Figure 3.15 **Sales area**

Figure 3.16 **Displays in the Maserati dealer are used to present a nice image**

Figure 3.17 **Cars on sale in a showroom**

important therefore to get this bit right. The reception should be manned by pleasant and qualified persons. The purpose of a reception and booking system within a company can be best explained by following through a typical enquiry. Your company may have a slightly different system but it will be similar.

- The customer enters reception area and is greeted in an appropriate way.
- Attention is given to the customer to find out what is required. (Let's assume the car is difficult to start, in this case.)
- Further questions can be used to determine the particular problem, bearing in mind the knowledge of vehicles the customer may, or may not have. (Is the problem worse when the weather is cold, for example?)
- Details are recorded on a job card about the customer, the vehicle and the nature of the problem. If the customer is new a record card can be started, or continued for an existing customer.
- Explanation of expected costs is given as appropriate. An agreement to only spend a set amount, after which the customer will be contacted, is a common and sensible approach.
- Date and time when the work will be carried out can now be agreed. This depends on workshop time availability and when is convenient for the customer. It is often better to say you cannot do the job until a certain time, rather than make a promise you can't keep.

- The customer is thanked for visiting. If the vehicle is to be left at that time, the keys should be labelled and stored securely.
- Details are now entered in the workshop diary or loading chart (usually computer based).

Parts department The parts department is the area where parts are kept and or ordered. This will vary quite a lot between different companies. Large main dealers will have a very large stock of parts for their range of vehicles. They will have a parts manager and in some cases several other staff. In some very small garages the parts department will be a few shelves where popular items such as filters and brake pads are kept.

Parts stock Even though the two examples given previously are rather different in scale the basic principles are the same and can be summed up very briefly as follows:

- A set level of parts or stock is decided upon.
- Parts are stored so they can be easily found.
- A re-ordering system should be used to maintain the stock.

Security Security is important as most parts cost a lot of money. When parts are collected from the parts department or area, they will be for use in one of three ways:

Figure 3.18 Typical parts department in a main dealer

Figure 3.19 **Typical parts department in an independent retailer**

Figure 3.20 **Tyres in stock**

Remember! There is a multimedia version of this textbook that includes additional images and interactive features: www.automotivett.org

- For direct sale to a customer.
- To be used as part of a job.
- For use on company vehicles.

In the first case an invoice or a bill will be produced. The second case, the parts will be entered on the customers job card. The third case may also have a job card or if not some other record must be kept. In all three cases keeping a record of parts used will allow them to be reordered if necessary. If parts are ordered and delivered by an external supplier, again they must be recorded on the customer's job card.

Figure 3.21 **Even the best cars need a wash**

Figure 3.22 **Smile, and enjoy your work!**

Figure 3.23 **Workshop**

Summary To operate a modern automotive business is a complex process. However, the systems outlined in this section have given an overview that shows how when each part of a complex system is examined, it is much easier to understand and appreciate the bigger picture.

Create an information wall to illustrate the important features of a component or system in this section.

Now complete the multiple choice quiz associated with this topic/ subject area.

3.2 EL13 Light Vehicle Construction

Together with the multimedia resources, this section is ideal material for students working towards the IMI Awards unit EL13, the general light vehicle aspects of City and Guilds entry level units and all other similar foundation or introduction level qualifications.

After successful completion of this section you will be able to show you have achieved these outcomes:

- Know about vehicle layouts and drive line configurations.
- Know about body types for a range of vehicles.
- Know the names of the main body parts found on light vehicles.

3.2.1 What you must know about vehicle construction

Always wear suitable PPE and use tools and equipment safely and correctly when doing practical work on any vehicle

Body types for a range of vehicles: saloon, estate, hatchback, coupe, convertible, MPV and 4x4

The engine and drive line configurations for a range of vehicles: front-engine, rear-engine, and mid-engine, front-wheel drive, rear-wheel drive, four-wheel drive

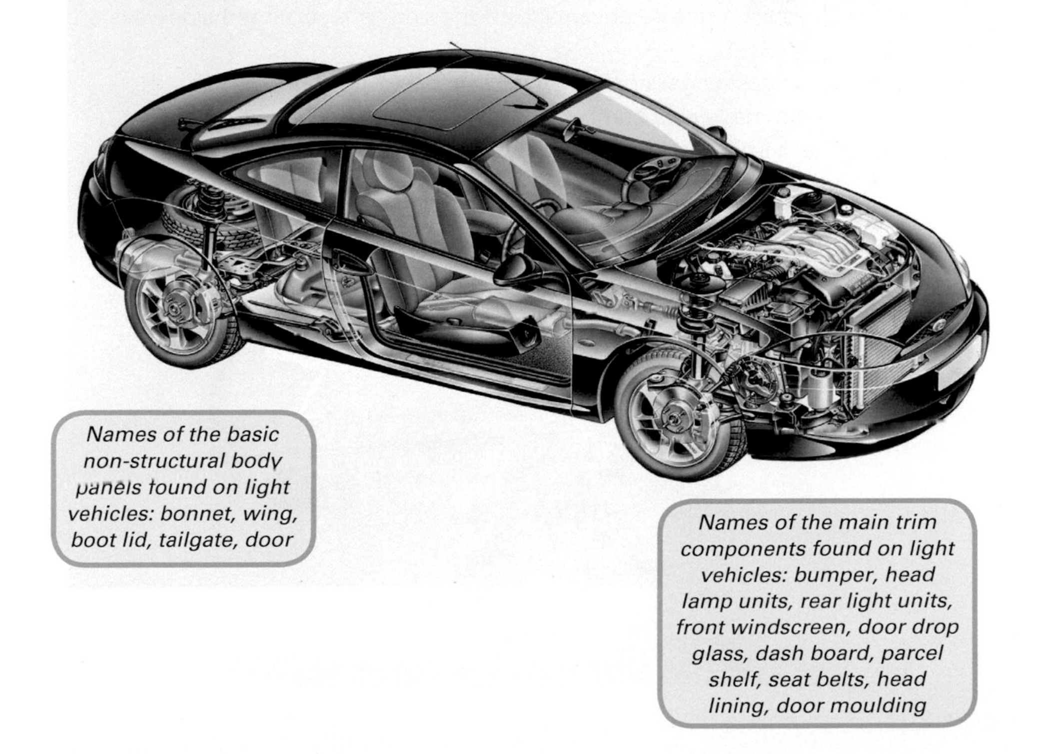

3.2.2 Layouts

This section is a general introduction to the car as a whole. Over the years many unusual designs have been tried, some with more success than others. The most common is of course a rectangular vehicle with a wheel at each corner! To take this rather simple idea further, we can categorize vehicles in different ways. For example, by layout such as:

- front-engine driving the front wheels
- front-engine driving the rear wheels
- front-engine driving all four wheels
- rear-engine driving the rear wheels
- mid-engine driving the rear wheels
- mid-engine driving all four wheels.

FWD	Front-wheel drive
RWD	Rear-wheel drive
AWD	All-wheel drive
4WD	Four-wheel drive

The following paragraphs and bullet points highlight features of the vehicle layouts mentioned above.

A common layout for a standard car is the front-engine, front-wheel drive vehicle. This is because a design with the engine at the front driving the front wheels has a number of advantages.

- Protection in case of a front-end collision.
- Easier engine cooling because of the air flow.
- Cornering can be better if the weight is at the front.
- Front-wheel drive adds further advantages if the engine is mounted sideways on (transversely).
- More room in the passenger compartment.
- Power unit can be made as a complete unit.
- Drive acts in the same direction that the steered wheels are pointing.

Figure 3.24 Front engine FWD

Front engine rear-wheel drive (RWD) from a front engine was the method used for many years. Some manufacturers have continued its use, BMW for example. A long propeller shaft from the gearbox to the final drive, which is part of the rear axle, is the main feature. The propshaft has universal joints to allow for suspension movement. This layout has advantages:

- weight transfers to the rear driving wheels when accelerating; and
- complicated constant velocity joints such as used by front-wheel drive vehicles, are not needed.

Four-wheel drive combines all the good points mentioned above but does make the vehicle more complicated and therefore expensive. The main difference with four-wheel drive is that an extra gearbox known as a transfer box is needed to link the front and rear-wheel drive.

Figure 3.25 Front engine RWD

The rear engine design has not been very popular but it was used for the bestselling car of all time – the VW Beetle. The advantages are that weight is placed on the rear wheels giving good grip and the power unit and drive can

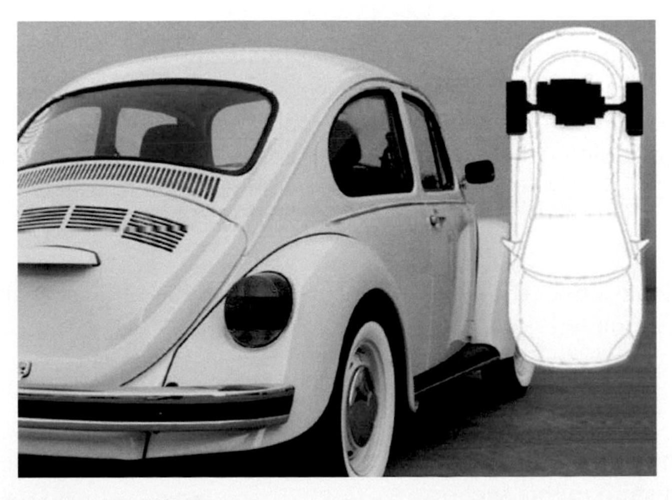

Figure 3.26 **Rear engine RWD**

be all one assembly. One down side is that less room is available for luggage in the front. The biggest problem is that handling is affected because of less weight on the steered wheels. Flat-type engines are the most common choice for this type of vehicle.

Fitting the engine in the mid position of a car has one major disadvantage; it takes up space inside the vehicle. This makes it impractical for most 'normal' vehicles. However, the distribution of weight is very good, which makes it the choice of high performance vehicle designers. A good example is the Ferrari Testarossa. Mid-engine is the term used to describe any vehicle where the engine is between the axles, even if it is not in the middle.

Figure 3.27 **Mid-engine RWD**

Vehicles are also categorized by type and size as in the following table.

LV	Light vehicles (light vans and cars) with a maximum allowed mass (MAM) of up to 3,500 kg, no more than eight passenger seats Vehicles weighing between 3,500 kg and 7,500 kg are considered as mid-sized.
LGV	A large goods vehicle, known formerly and still in common use, as heavy goods vehicle or HGV. LGV is the EU term for trucks or lorries with a MAM of over 3,500 kg.
PCV	A passenger carrying vehicle or a bus, known formerly as omnibus, multibus or autobus) is a road vehicle designed to carry passengers. The most common type is the single-decker, with larger loads carried by double-decker and articulated buses, and smaller loads carried by minibuses. A luxury, long distance bus is usually called a coach.

Look back over the previous section and write out a list of the key bullet points.

Create a mind map to illustrate the important features of a component or system in this section.

3.2.3 Body design

Figure 3.28 **Saloon car (Source: Ford Media)**

Figure 3.29 **Estate car (Source: Ford Media)**

Figure 3.30 **Hatchback (Source: Ford Media)**

Figure 3.31 **Coupe (Source: Ford Media)**

Figure 3.32 **Convertible (Source: Ford Media)**

Figure 3.33 **Concept car (Source: Ford Media)**

Types of light vehicle can range from small two-seat sports cars to large people carriers or SUVs. Also included in the range are light commercial vehicles such as vans and pick-up trucks. It is hard to categorize a car exactly as there are several agreed systems in several different countries. Figures 3.28–3.36 show a number of different body types.

The vehicle chassis can be of two main types: separate or integrated. Separate chassis are usually used on heavier vehicles. The integrated type, often called monocoque, is used for almost all cars. The two main types are shown here as Figures 3.37 and 3.38.

Figure 3.34 **Light van (Source: Ford Media)**

Figure 3.35 **Pickup truck (Source: Ford Media)**

Figure 3.36 **Sports utility vehicle (SUV) (Source: Ford Media)**

Figure 3.37 **Ladder chassis**

Figure 3.38 **Integrated chassis**

Most vehicles are made of a number of separate panels. Figure 3.39 shows a car with the main panel or other body component named.

Figure 3.39 **Body components**

Use the interactive media search tools to look for pictures and videos to examine the subject in this section in more detail.

Look back over the previous section and write out a list of the key bullet points.

3.3 Automotive industry puzzles

3.3.1 Cryptograms

A	B	C	D	E	F	G	H	I	J	K	L	M	N	O	P	Q	R	S	T	U	V	W	X	Y	Z
	20			24				19														7			

B	E		I	_	_	E	_	E	_	_	E	_		I	_		_	_	E		_	_	_	_	_	_	_
20	24		19	12	6	24	26	24	14	6	24	22		19	12		6	4	24		10	16	17	1	5	12	11

_	_	_		W	_	_	_		_	_	_
11	16	25		7	16	26	9		23	16	26

Cryptogram 3.1 Get involved:

3.3.2 Anagrams

A Training Zoo A Train Oozing

A Ratio Zoning Angora Ion Zit

Anagrams 3.1 Structure of your workplace: what word did I use to get these anagrams?

3.3.3 Word search

```
S J H Q U I V E Z P C R T W E
F T N X J W A D R L U Z J A N
C Y N A P M O C W X S F W D C
R O R E W A R R A N T Y P E M
I I N W M K B I W E O F S G S
E V S T C T N Z M B M G S A P
R G C O R V R O W P E T E R B
Y E T E O A O A H B R C N A D
Y S L I O R C N P K A D I G T
F J C A W S F T G E Z C S A N
N E N O E N T R U U D Z U D I
U J H E T D B H O L Q Q B K R
G S K R E R U T C A F U N A M
L O P O S U A C C E L G M P N
H A H W E O F Z U L T V A Q U
```

BUSINESS
COMPANY
CONTRACT
CUSTOMER
DEALER
DEPARTMENTS
GARAGE
INVOICE
MANUFACTURER
SHOWROOM
STOCK
WARRANTY

3.3.4 Crosswords

Crossword 3.1 Automotive industry

Across

1 What is needed to do a repair correctly

2 Passing information from one person to another

3 What is needed to keep people informed

4 A general description of our industry

5 The most important person in our business

Down

1 Paperwork or computer records

CHAPTER 4

Workshop Skills

4.1 Introduction to workshop tools and equipment

Together with the multimedia resources, this section is ideal material for students working towards the IMI Awards unit EL03, the City and Guilds 3902–012 unit and all other similar foundation or introduction level qualifications.

After successful completion of this section you will be able to show you have achieved these objectives:

- Know common motor vehicle hand tools and work shop equipment.
- Be able to use motor vehicle hand tools and workshop equipment correctly and safely.
- Know examples of measuring equipment used in a motor vehicle workshop.
- Know examples of different locking and securing devices used on motor vehicles.

Automotive Technician Training: Entry Level 3. 978-0-415-72040-3.
© Tom Denton Published by Taylor & Francis. All rights reserved.

4.1.1 What you must know about workshop tools and equipment:

Identify and demonstrate how to use these tools safely: spanners (open end, ring, combination, speed and ratchet types), screwdrivers (blade, Phillips, pozidrive), hammers (ball pein, lump, copper/hide, rubber, neoprene), Allen keys, vice (mole) grips, socket sets (different drive sizes for specific equipment to cover range of vehicles), pliers (long nose, engineers, side snips/ cutters), torque wrench

Identify and demonstrate the safe use of common equipment found in a motor vehicle workshop: lifting equipment (jacks, ramps and axle stands), air lines and attachments (wrenches, blow guns, tyre inflator/gauge), mains electrical apparatus (drills, extension leads, parts cleaner), task specific specialist tools (optical tracking gauges), filter straps, waste oil drainers

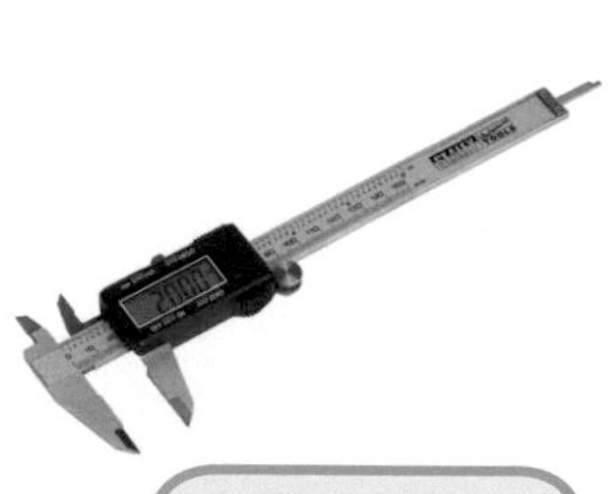

Name the basic measuring equipment used in the industry: tape measure, steel rule, feeler blades, vernier calipers, micrometers, tread depth gauges

Demonstrate the correct way to use locking devices: castellated nut with split pins, lock tabs, self -locking nuts, locking nuts (2 nut method), chemical lock

4.1.2 Hand tools

Figure 4.1 Ring spanners (wrenches)

Using hand tools is something you will learn by experience, but an important first step is to understand the purpose of the common types. This section therefore starts by listing some of the more popular tools, with examples of their use, and ends with some general advice and instructions.

Practise until you understand the use and purpose of the following tools when working on vehicles:

Hand tool	Example uses and/or notes
Adjustable spanner (wrench)	An ideal stand by tool and useful for holding one end of a nut and bolt.
Open-ended spanner	Use for nuts and bolts where access is limited or a ring spanner can't be used.
Ring spanner	The best tool for holding hexagon bolts or nuts. If fitted correctly it will not slip and damage both you and the bolt head.
Torque wrench	Essential for correct tightening of fixings. The wrench can be set in most cases to 'click' when the required torque has been reached. Many fitters think it is clever not to use a torque wrench. Good technicians realise the benefits.
Socket wrench	Often contain a ratchet to make operation far easier.
Hexagon socket spanner	Sockets are ideal for many jobs where a spanner can't be used. In many cases a socket is quicker and easier than a spanner. Extensions and swivel joints are also available to help reach that awkward bolt.
Air wrench	These are often referred to as wheel guns. Air driven tools are great for speeding up your work but it is easy to damage components because an air wrench is very powerful. Only special, extra strong, high-quality sockets should be used.
Blade (engineer's) screwdriver	Simple common screw heads. Use the correct size!

Remember! There is a multimedia version of this textbook that includes additional images and interactive features: www.automotivett.org

Hand tool	Example uses and/or notes
Pozidrive, Philips and cross-head screwdrivers	Better grip is possible particularly with the Pozidrive but learn not to confuse the two very similar types. The wrong type will slip and damage will occur.
Torx®	Similar to a hexagon tool like an Allen key but with further flutes cut in the side. It can transmit good torque.
Special purpose wrenches	Many different types are available. As an example mole grips are very useful tools as they hold like pliers but can lock in position.
Pliers	These are used for gripping and pulling or bending. They are available in a wide variety of sizes. These range from snipe nose, for electrical work, to engineers pliers for larger jobs such as fitting split pins.
Levers	Used to apply a very large force to a small area. If you remember this you will realise how if incorrectly applied, it is easy to damage a component.
Hammer	Anybody can hit something with a hammer, but exactly how hard and where is a great skill to learn!

General advice and instructions for the use of hand tools (taken from information provided by Snap-on):

- Only use a tool for its intended purpose.
- Always use the correct size tool for the job you are doing.
- Pull a spanner or wrench rather than pushing whenever possible.
- Do not use a file or similar, without a handle.
- Keep all tools clean and replace them in a suitable box or cabinet.
- Do not use a screwdriver as a pry bar.
- Look after your tools and they will look after you!

Note The following screens give an overview of some of the tools you will need to become familiar with – there are many more! As you study each main chapter it may be useful to refer back to the tools and equipment presented here.

Hose removal A number of tools have been developed to aid the removal of hoses. These cranked and blunt-bladed probes can be eased into the end of a hose to break the seal against the connector pipe. Blunt screwdrivers can also perform this task. Another method is to use a sharp knife to cut the hose back, but this method can only be used when the hose is to be discarded.

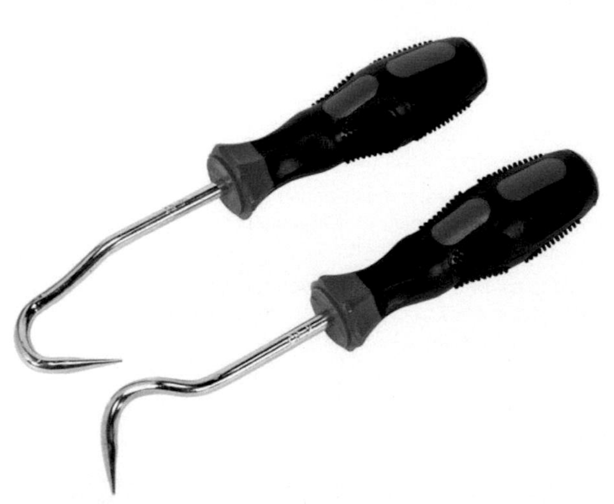

Figure 4.2 Hose remover

Hose clips There are a range of specialised hose clips that require a dedicated pair of pliers for their removal and installation. These tend to be specific to individual manufacturers, and the pliers are included in the workshop-equipment pack for the dealer. Where the special tools are not available, the clip can be replaced with a general-purpose screw-type hose clip.

Strap wrenches For removing and replacing fuel and oil canister-type filters.

Figure 4.3 Strap wrench

Tools for exhaust removal Some useful tools for exhaust-system removal are chain wrenches for twisting seized pipes, and oxyacetylene welding equipment for freeing up rusted joints. An air chisel or cutter may be necessary for cutting off components that will not be reused.

Figure 4.4 Chain cutter

Figure 4.5 Air cutter

Torque wrench A good torque wrench is an essential piece of equipment. Many types are available but all work on a similar principle. Most are set by adjusting a screwed cylinder, which forms part of the handle. An important point to remember is that, as with any measuring tool, regular calibration is essential to ensure it remains accurate.

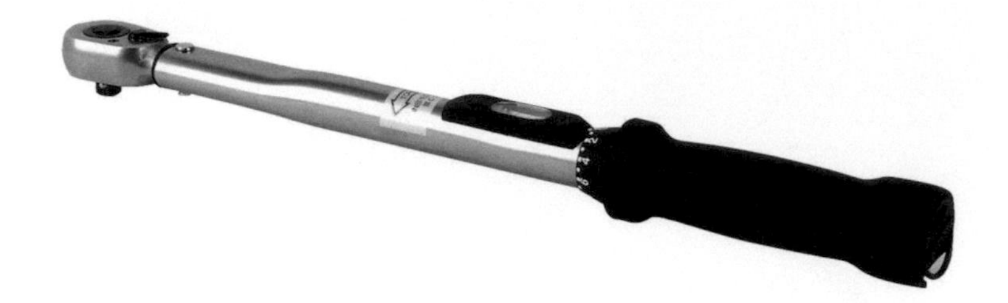

Figure 4.6 Torque wrench

Bearing puller Removing some bearings is difficult without a proper puller. For internal bearings, the tool has small legs and feet that hook under the bearing. A threaded section is tightened to pull out the bearing. External pullers hook over the outside of the bearing and a screwed thread is tightened against the shaft.

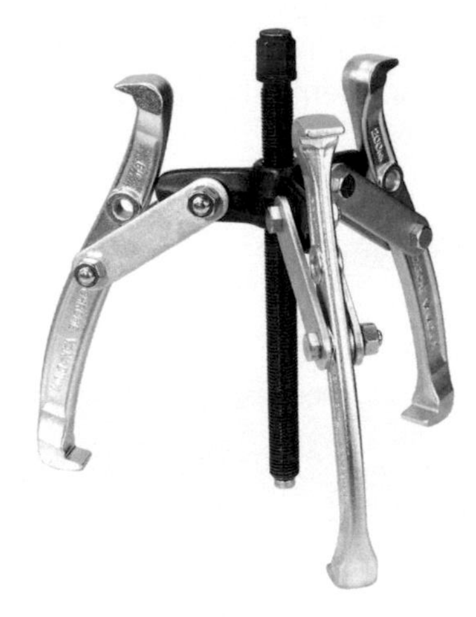

Figure 4.7 External bearing puller

Soft hammers These tools allow a hard blow without causing damage. They are ideal for working on driveshafts, gearboxes and final drive components. Some types are made of special hard plastics whereas some are described as copper/hide mallets. This type has a copper insert on one side and a hide or leather insert on the other. It is still possible to cause damage, however, so you must take care!

Brake adjusting tools On many earlier braking systems, the adjustment (gap between the shoe and drum) had to be adjusted manually during a service. Most modern systems do this automatically. However, many earlier systems are still

Figure 4.8 Some hammers contain metal shot to give a 'dead blow'

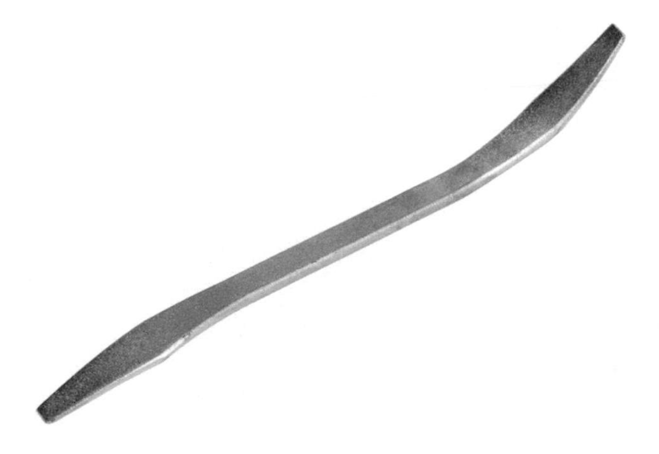

Figure 4.9 These tools are simple levers

in use so tools such as these, which are used to rotate a gear inside the drum, will be very useful. Some are made to suit particular manufacturer's systems.

Test lamp This is an often underrated piece of test equipment! However, it must be used with care. The advantage of a simple test lamp is that it draws some current through the circuit under test. This allows 'high resistance' faults to be located easily. However, this is also a disadvantage because drawing current through electronic circuits can damage them. Using a test lamp for checking supplies to electrical items such as lights and motors is fine; for all other tests, a multimeter is the preferred option. If in doubt, consult manufacturers' data.

Use the interactive media search tools to look for pictures and videos to examine the subject in this section in more detail.

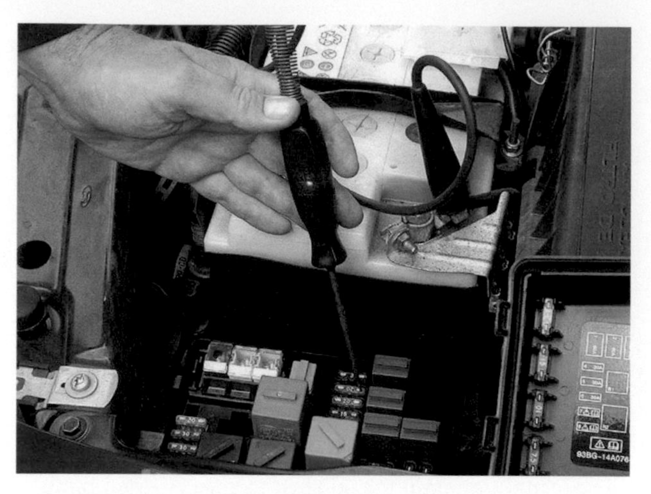

Figure 4.10 Test lamps are simple but useful

Jumper wire A jumper wire is useful for bypassing components such as switches. However, do not short supplies to earth using this method. As a safety feature, it is recommended that the jumper wire be fitted with a fuse. A value of 5 to 10A is probably ideal. Crocodile clips or spade terminal ends can be very useful for testing purposes.

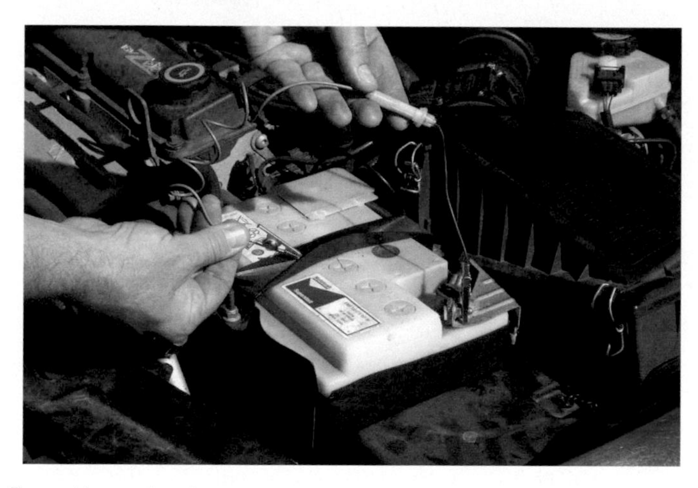

Figure 4.11 Fused jump lead

Terminal Kit Many terminal kits are available. They usually consist of a selection of terminals and special pliers to crimp the terminals on to the wire.

Wire Strippers With practice you will be able to strip wire using side cutters. However, special tools are available to make the job easier. A number of different types are shown here.

Soldering iron Most soldering irons are electrically heated. However, there are some very good gas powered types now available. The secret with a soldering iron is to use the right size for a specific job. One suitable for delicate ICs and circuit boards will not work on large alternator diodes. More damaging would be to use a large iron on a small circuit board!

Figure 4.12 Selection of wire strippers

Paper clip Not found in Snap-on or other catalogues, but a very useful tool. It is not only ideal for bridging terminals as shown here; it can also be used for clipping paper together!

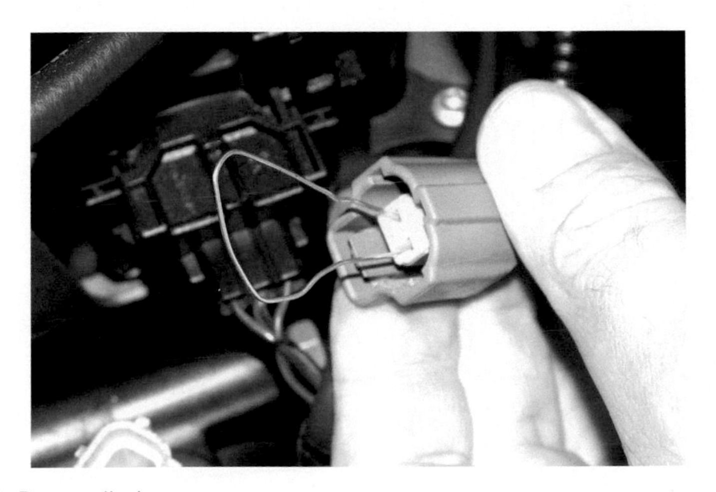

Figure 4.13 Paper clip in use

4.1.3 Measurement

Introduction Measurement is the act of measuring or the process of being measured. It is the process of comparing a dimension, quantity or capacity to a known value by using a suitable instrument. Simple examples of measuring instruments would be a tape measure or a steel rule (often referred to as a ruler).

Accuracy is a key aspect of measurement and generally means how close the measured value of something is, to its actual value. For example, if a length of about 30cm is measured with an old tape measure, then the reading may be 1 or 2 mm too high or low. This would be an accuracy of ± 2 mm.

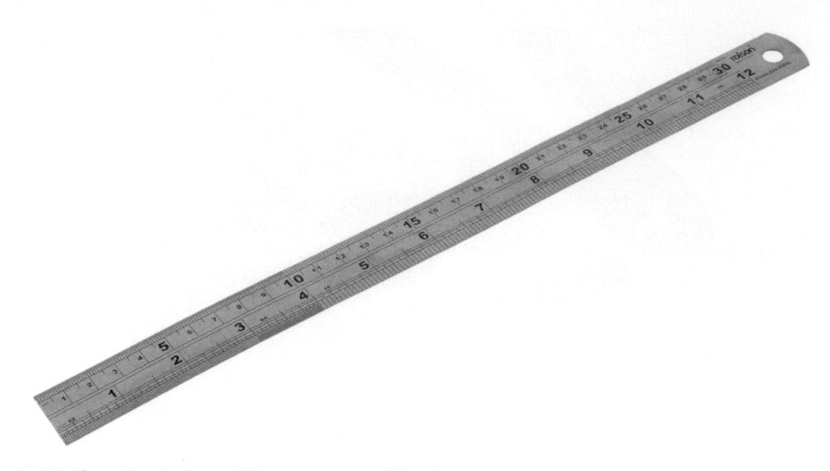

Figure 4.14 Steel rule in millimetres and inches

Now consider measuring a length of metal bar with a steel rule. How accurately could you measure it this time? Probably to the nearest 0.5mm if you were careful but this raises a number of issues to consider:

1. You could make an error reading the rule.
2. Do we actually need to know the length of a piece of bar to the nearest 0.5mm?
3. The rule may be damaged and not give the correct reading!

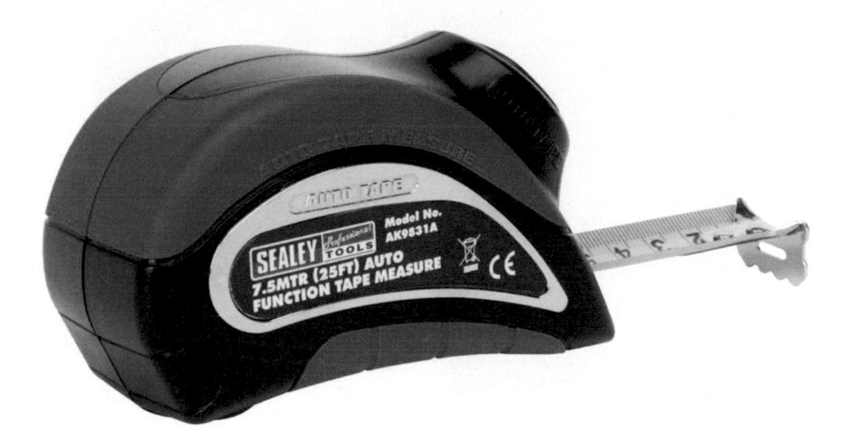

Figure 4.15 Tape measure (Source: Jack Sealey)

The first and second of these issues can be dispensed with by knowing how to read the test equipment correctly and also knowing the appropriate level of accuracy required. A micrometer for a tyre tread depth would be over the top, and in the same way, a tape measure for valve clearances would be nowhere near good enough. I am sure you get the idea.

Equipment To ensure instruments are, and remain accurate, there are just two simple guidelines:

- Look after the equipment, a micrometer thrown on the floor will not be accurate.
- Ensure instruments are calibrated regularly – this means being checked against known good equipment.

Here is a summary of the steps to ensure a measurement is accurate:

Step	Example
Decide on the level of accuracy required.	Do we need to know that the battery voltage is 12.6V or 12.635V?
Choose the correct instrument for the job.	A micrometer to measure the thickness of a shim
Ensure the instrument has been looked after and calibrated when necessary.	Most instruments will go out of adjustment after a time. You should arrange for adjustment at regular intervals. Most tool suppliers will offer the service or in some cases you can compare older equipment to new stock
Study the instructions for the instrument in use and take the reading with care. Ask yourself if the reading is about what you expected.	Is the piston diameter 70.75mm or 170.75mm?
Make a note if you are taking several readings	Don't take a chance: write it down

Straight edge and feelers A 'straight edge' is a piece of equipment with a straight edge! It is used as a reference for measuring flatness. It is placed on top of the test subject. The feeler blades are then used to assess the size of any gaps. The feeler blades are sized in either hundredths of a millimetre or thousandths of an inch.

Figure 4.16 Straight edge

Figure 4.17 Feeler gauges

Micrometer The most common uses for a micrometer in the motor trade are for measuring valve shims, brake disc thickness and crankshaft journals.

The spindle of an ordinary metric micrometer has 2 threads per millimetre, and therefore one complete revolution moves the spindle through a distance of 0.5 mm. The longitudinal line on the frame is graduated with 1 mm divisions and 0.5 mm subdivisions. The thimble has 50 graduations, each being 0.01 mm (one-hundredth of a millimetre). Thus, the reading is given by the number of millimetre divisions visible on the scale of the sleeve plus the particular division on the thimble which coincides with the axial line on the sleeve.

Example Suppose that the thimble were screwed out so that graduation 5, and one additional 0.5 subdivision were visible (as shown in the image), and that graduation 28 on the thimble coincided with the

Figure 4.18 Zero to 25mm micrometer

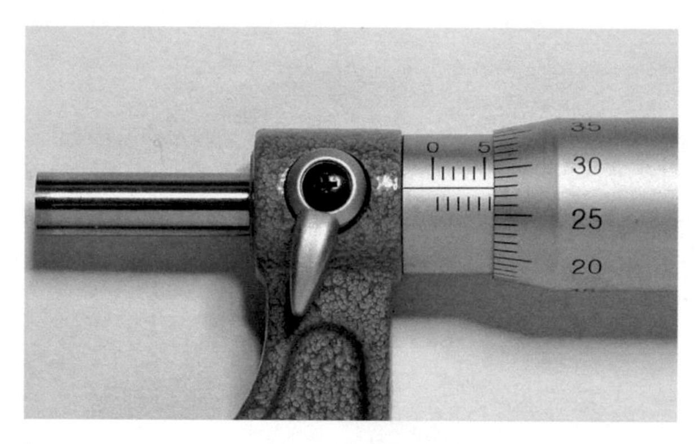

Figure 4.19 Metric system – micrometer thimble reading 5.78mm

axial line on the sleeve. The reading then would be 5.00 + 0.5 + 0.28 = 5.78 mm.

Dial gauge A dial gauge is a device used to test the movement of something very accurately. A good example is the run out of brake discs. A plunger on the dial gauge is made to run up against the disc as it is turned. The body of the gauge is clamped in position and via accurate gears the movement of the plunger makes a hand turn on a clock dial. The face of the clock is

Figure 4.20 Dial gauge

marked off in 0.01mm increments and will usually rotate enough times to measure up to 10mm. Diesel pump timing is often set or checked with a dial gauge. The pumping plunger is made to act on the dial gauge so that as the engine is turned it is possible to tell the position of the injection plunger very accurately.

Vernier caliper There are now three types of caliper used for accurate measuring: vernier, dial and digital. These calipers have a calibrated scale with a fixed jaw, and another jaw, with a pointor, that slides along the scale. The distance between the jaws is then read in different ways for the three types. Vernier, dial and digital calipers can measure internal dimensions, external dimensions using the lower jaws, and in many cases depth by the use of a probe that is attached to the movable head and slides along the centre of the body. This probe is slender and can get into deep grooves that may prove difficult for other measuring tools. The vernier scales may include metric measurements on the lower part of the scale and inch measurements on the upper, or vice versa, in countries that use inches. Vernier calipers commonly used in industry provide a precision to 0.01 mm, or 0.001 inch. They are available in various sizes.

Figure 4.21 Digital caliper

Reading To take a reading on the digital type is simple as it is displayed as a figure. The dial type displays the major distance on the sliding scale and then greater accuracy is achieved by adding the reading on the dial. The vernier type (as shown here) is read as follows: First read the position of the pointer directly on the millimetre scale (24 mm in this case). When the pointer is between two markings (as it is here), you can mentally interpolate to improve the precision of the reading (my guess would be about 24.8). However, the addition of the vernier scale allows a more accurate interpolation. All you need to do is check which mark on the vernier scale lines up exactly with one of the main scale. In this example I would say it is the one half way between the 7 and the 8. This means the actual measurement is 24.75 mm.

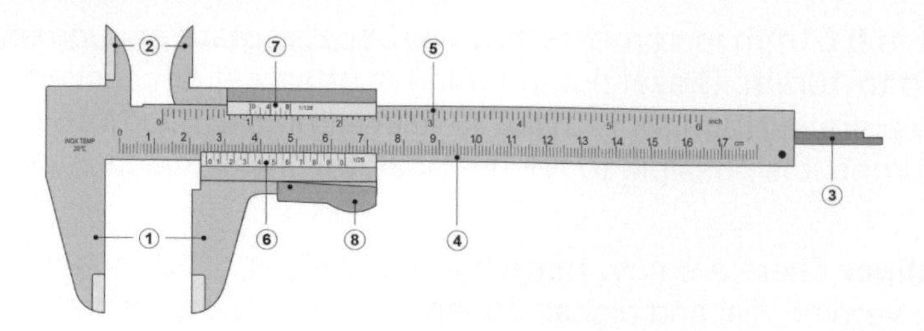

Figure 4.22 Vernier caliper – the main parts are: 1. Outside jaws: used to measure external diameter or width of an object, 2. Inside jaws: used to measure internal diameter of an object, 3. Depth probe: used to measure depths of an object or a hole, 4. Main scale: scale marked every mm, 5. Main scale: scale marked in inches and fractions, 6. Vernier scale gives interpolated measurements to 0.01 mm or better, 7. Vernier scale gives interpolated measurements in fractions of an inch, 8. Retainer: used to block movable part to allow the easy transferring of a measurement

Angle locator This magnetic device is used to check that the angles of a propshaft are equal. This is important because it ensures that the changing velocity effects of the universal joints (UJs) are cancelled out. The angle locator attaches magnetically to the shaft. A dial is set to zero and then, when it is moved to a new location, the difference in angle is indicated.

Figure 4.23 This device checks propshaft angles

Multimeter An essential tool for working on vehicle electrical and electronic systems is a good digital multimeter (often referred to as a DMM). Digital meters are most suitable for accuracy of reading as well as other facilities.

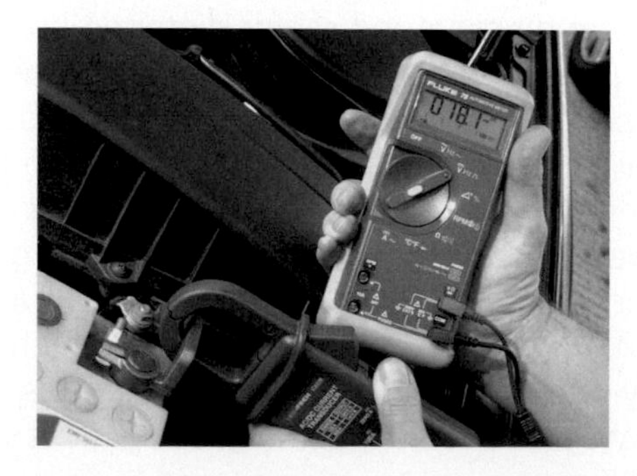

Figure 4.24 Multimeter and current clamp

The following list of functions broadly in order, starting from essential to desirable should be considered:

Function	Range	Accuracy
DC voltage	500V	0.3%
DC current	10A	1.0%
Resistance	0 to 10MW	0.5%
AC voltage	500V	2.5%
AC current	10A	2.5%
Dwell	3,4,5,6,8 cylinders	2.0%
rpm	10,000 rpm	0.2%
Duty cycle	% on/off	0.2% /kHz
Frequency	over 100kHz	0.01%
Temperature	> 9000C	0.3% +30C
High current clamp	1000A (DC)	Depends on conditions
Pressure	3 bar	10.0% of standard scale

Oscilloscope An oscilloscope is a very useful measuring instrument. There were traditionally two types of oscilloscope; analog or digital. However, the digital scope is now universal. An oscilloscope draws a graph of voltage (the vertical scale or Y axis) against time (the horizontal scale or X axis).The trace is made to move across the screen from left to right and then to 'fly back' and start again. The frequency at which the trace moves across the screen is known as the time base, which can be adjusted either automatically or manually. The signal from the item under test can either be amplified or attenuated (reduced), much like changing the scale on a voltmeter.

The trigger, which is what starts the trace moving across the screen starts, can be caused internally or externally. When looking at signals such as ignition voltages, triggering is often external, each time an individual spark fires or each time number one spark plug fires for example.

PicoScope The Pico automotive kit turns a laptop or desktop PC into a powerful automotive diagnostic tool for fault finding sensors, actuators and electronic circuits. The oscilloscope connects to a USB port on a PC and can

Figure 4.25 Automotive oscilloscope kit (Source: PicoTech)

Remember! There is a multimedia version of this textbook that includes additional images and interactive features: www.automotivett.org

take up to 32 million samples per trace, making it possible to capture complex automotive waveforms. The scope can be used to measure and test virtually all of the electrical and electronic components and circuits in any modern vehicle including:

- Ignition (primary and secondary).
- Injectors and fuel pumps.
- Starter and charging circuits.
- Batteries, alternators and starter motors.
- Lambda, Airflow, knock and MAP sensors.
- Glow plugs / timer relays.
- CAN bus, LIN bus and FlexRay.

Waveform When you look at a waveform on a screen it is important to remember that the height of the scale represents voltage and the width represents time. Both of these axes can have their scales changed. They are called axis because the 'scope' is drawing a graph of the voltage at the test points over a period of time. The time scale can vary from a few ms to several seconds. The voltage scale can vary from a few mV to several kV. For most test measurements only two connections are needed just like a voltmeter.

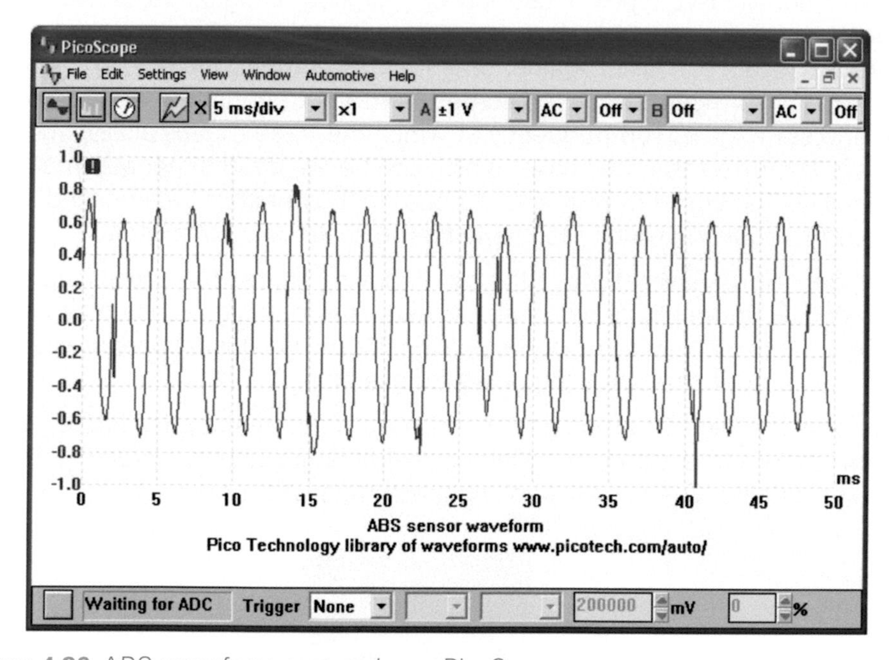

Figure 4.26 ABS waveform captured on a PicoScope

Use the interactive media search tools to look for pictures and videos to examine the subject in this section in more detail.

Now complete the multiple choice quiz associated with this topic/ subject area.

4.1.4 Workshop equipment

In addition to hand tools and test equipment, most workshops will also have a range of equipment for lifting and supporting as well as electrical or air operated tools. The table lists some examples of common workshop equipment together with typical uses.

Equipment	Common use
Ramp or hoist	Used for raising a vehicle off the floor. Figure 4.27 shows a two-post wheel-free type. Other designs include four-post and scissor types where the mechanism is built in to the workshop floor
Jack and axle stands	A trolley jack (such as Figure 4.28) is used for raising part of a vehicle such as the front or one corner or side. It should always be positioned under suitable jacking points, axle or suspension mountings. When raised, stands must always be used in case the seals in the jack fail causing the vehicle to drop
Air gun	A high pressure air supply is common in most workshops. Figure 4.30 is a typical wheel gun used for removing wheel nuts or bolts. Note that when replacing wheel fixings it is essential to use a torque wrench
Electric drill	The electric drill is just one example of electric power tools used for automotive repair. Note that it should never be used in wet or damp conditions
Parts washer	There are a number of companies that supply a parts washer and change the fluid it contains at regular intervals.
Steam cleaner	Steam cleaners can be used to remove protective wax from new vehicles as well as to clean grease, oil and road deposits from cars in use. They are supplied with electricity, water and a fuel to run a heater – so caution is necessary
Electric welder	There are a number of forms of welding used in repair shops. The two most common are metal inert gas (MIG) and manual metal arc (MMA). Figure 4.29 shows the MMA process.
Gas welder	Gas welders are popular in workshops as they can also be used as a general source of heat, for example, when heating a flywheel ring gear
Engine crane	A crane of some type is essential for removing the engine on most vehicles. It usually consists of two legs with wheels that go under the front of the car and a jib that is operated by a hydraulic ram. Chains or straps are used to connect to or wrap around the engine
Transmission jack	On many vehicles the transmission is removed from underneath. The car is supported on a lift (perhaps similar to Figure 4.27) and then the transmission jack is rolled underneath. An example is shown as Figure 4.31

Figure 4.27 Car lift (Source: asedeals.com)

Figure 4.28 Trolley jack and axle stands (Source: Snap-on Tools)

Figure 4.29 Welding process (Source: Wikimedia)

Note The following screens give an overview of some of the equipment you will need to become familiar with – there is a lot more! As you study each main chapter it may be useful to refer back to the tools and equipment presented here.

AC servicing unit Most modern servicing units can be used to drain, recycle, evacuate and refill air conditioning systems. Some older types would only carry out individual procedures. Note that different servicing units are required for R12 and R134a refrigerants and their oils must not be mixed with one another. If work is to be carried out on an air conditioning system,

Figure 4.30 Wheel gun (Source: Snap-on Tools)

Figure 4.31 Transmission jack (Source: Snap-on Tools)

a servicing unit is essential. Refrigerant must never be released into the atmosphere.

Coil spring compressors Springs must be compressed before they are removed from suspension struts. A number of different tools are available. However, the type shown here is very popular. The two clamps are positioned either side of the spring using the hooked ends. The bolts are then tightened evenly until the tension of the spring is taken by the clamps.

Ball joint presses To remove a taper fitting ball joint, a splitter or press is usually needed. If the joint is to be reused, a lever or clamp-type splitter is

Figure 4.32 AC Servicing equipment

Figure 4.33 McPherson strut springs must be compressed for removal

preferred. The tool clamps onto the arm and threaded section of the joint. A bolt is tightened, which applies a force to push the joint free.

Ball joint splitter Two types of ball joint splitter are in common use. One type is a simple forked wedge that is hammered in between the joint and the arm. This works well but can damage the joint. If the joint is to be reused, the lever-type splitter is preferred. This tool clamps onto the arm and threaded section of the joint. A bolt is tightened, which applies a force to push the joint free.

Figure 4.34 This clamp removes tapered ball joints

Figure 4.35 Lever-type splitter

Pipe clamp A pipe clamp is used to block a pipe for tests or repairs to be carried out. For example, on a braking system, it can be used to prevent leakage of fluid when cylinders are replaced. Alternatively, the source of spongy brakes can be narrowed down. This is done by clamping each flexible pipe in turn and pressing the pedal. However, some manufacturers do not recommend these tools because the pipe can be damaged.

Pressure bleeder This equipment forces fluid through the reservoir under pressure. The tank is in two parts, separated by a diaphragm. The top of the tank is filled with new brake fluid, and the lower part pressurised with compressed air. Using suitable adaptors, the outlet pipe, from the fluid section, is connected to the master cylinder reservoir. A valve is opened and fluid is force out of the slave cylinders as the bleed nipples are opened. Fluid

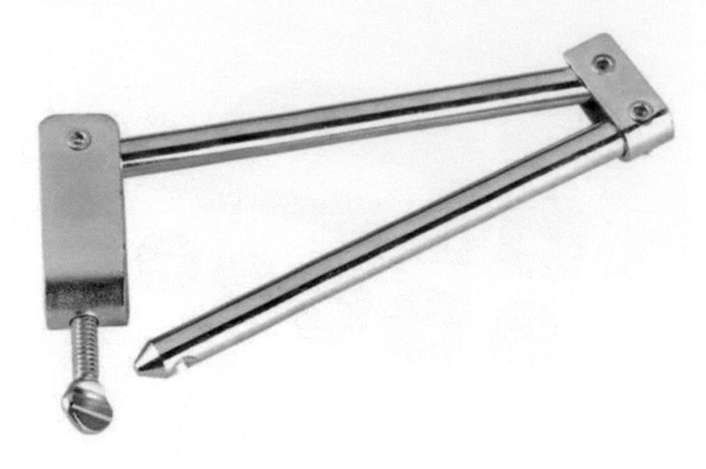

Figure 4.36 Only use recommended types

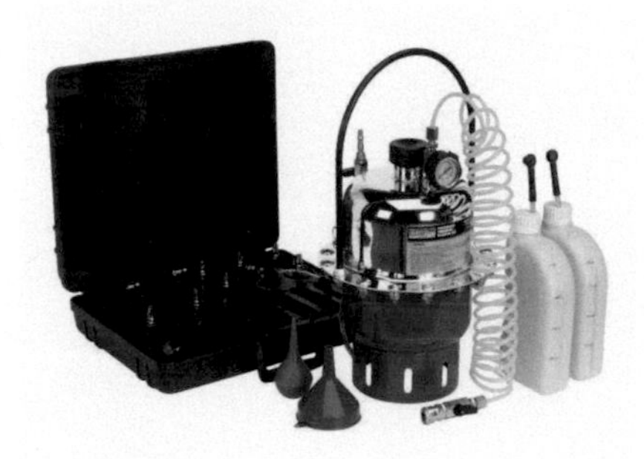

Figure 4.37 This equipment forces fluid through the reservoir

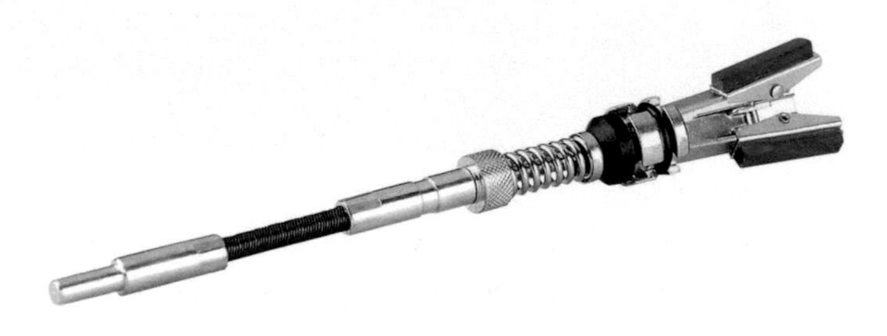

Figure 4.38 Engine cylinder honing tool

is collected in a container using a simple rubber pipe, just like when bleeding the system manually.

Honing tool A honing tool is sometimes called a 'glaze buster'. It is used to grind the inside of a cylinder to a good, final finish. This can be done to an engine cylinder or a much smaller hydraulic brake cylinder. The tool is usually mounted in an air drill as the power source. Lubrication should be used when the equipment is operated.

Figure 4.39 Balancer (Source: Bosch Media)

Wheel balancer Most wheel balancers offer facilities for measuring the wheel, and then programming this into a computer. The machines usually run from a mains electrical supply. The wheel is clamped to the machine and spun. Sensors in the machine determine the static and dynamic balance. A display states where extra weights should be added to obtain accurate balance. 'On-car' balancers have been used, but are less accurate than the later computerised types.

Tyre changer It is possible to change tyres with two levers and a hammer! However, it is much quicker and easier with an automatic changer. A lever is still needed to start the bead of the tyre lifting over the rim. An electric motor drives the wheel round as the tyre is removed or fitted. Most changers incorporate a bead breaker.

Tyre inflator This is a simple but important item of equipment. Make sure it is looked after so that the gauge remains accurate. A small difference in tyre pressure can have a significant effect on performance and wear.

Clutch aligner kit The clutch disc must be aligned with the cover and flywheel when it is fitted. If not, it is almost impossible, on some vehicles to replace the gearbox. This is because the gearbox shaft has to fit through the disc and into the pilot or spigot bearing in the flywheel. The kit shown here has adaptors to suit most vehicles.

Figure 4.40 Many types of changer are available (Source: Bosch Media)

Figure 4.41 The gauge must be accurate

Pilot/spigot bearing puller Removing spigot bearings is difficult without a proper puller. This tool has small legs and feet that hook under the bearing. A threaded section is tightened to pull out the bearing.

Slide hammer A slide hammer is a form of puller. It consists of a steel rod over which a heavy mass slides. The mass is 'hammered' against a stop, thus applying a pulling action. The clamp end of the tool can screw either into, or onto, the component. Alternatively, puller legs with feet are used to grip under the sides of the component.

Figure 4.42 The clutch must be aligned when fitted

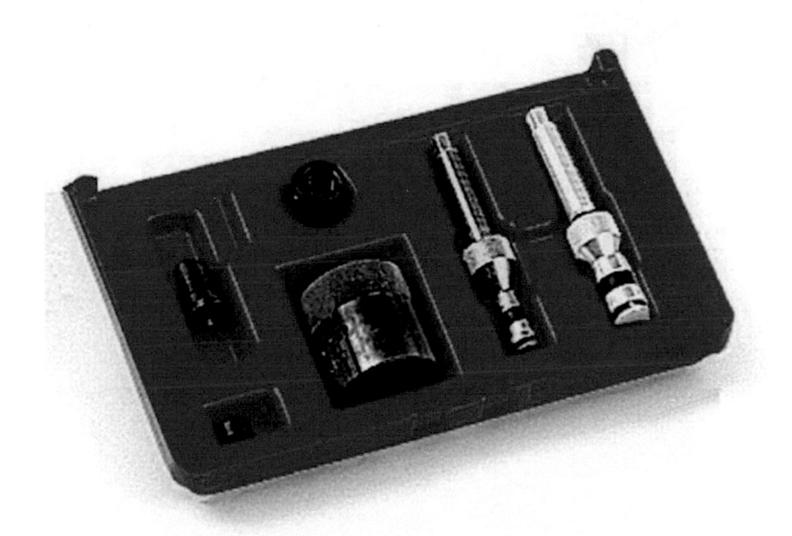

Figure 4.43 An internal bearing puller

4.1.5 Nuts, screws, washers and bolts

Introduction A fastener is a hardware device that mechanically joins or fixes two or more objects together. There is a large number of fasteners used on automobiles, some standard and some specialist. Most of the main types are outlined in this section. Three major types of steel are used for fasteners in the automotive and other similar industries; they are stainless steel, carbon steel and alloy steel. Nuts and bolts and screws and variations on the theme

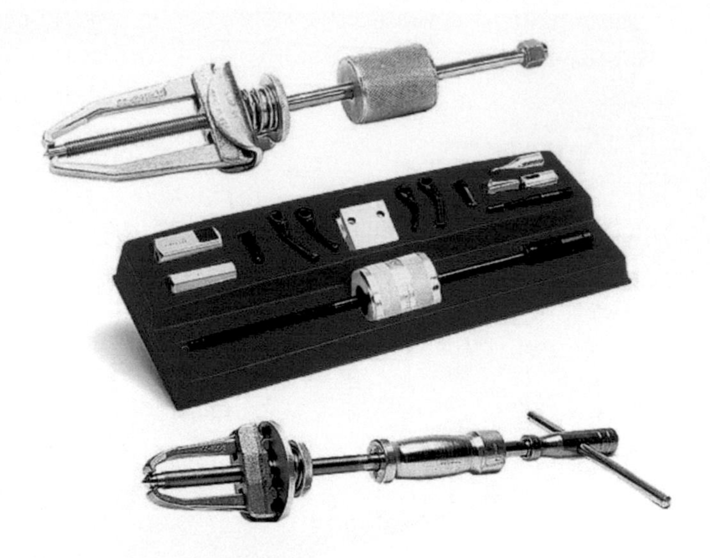

Figure 4.44 This tool is useful for removing halfshafts

Figure 4.45 A selection of fasteners

Figure 4.46 A few spare nuts in case you drop one. . .

are the most common type of fastener. They can also be described as a non-permanent fixing method (because they can be undone!).

Nuts A nut is a type of fastener with a threaded hole. Nuts are almost always used with a corresponding bolt or stud to fasten two or more parts together. The nut and bolt are kept together by a combination of their threads' friction, a slight stretch of the bolt, and compression of the parts. In applications where vibration or rotation may work a nut loose, various locking mechanisms are used (discussed later).

Locking nuts Adhesives, safety pins or lock wire, nylon inserts or slightly oval-shaped threads are used for this purpose. The most common shape for a nut is hexagonal, for similar reasons as the bolt head – six sides give a good range of angles for a tool to approach in tight spots, but more (and smaller) corners would be vulnerable to being rounded off. Other specialised shapes

Figure 4.47 These bolts are holding Clifton Suspension Bridge together (designed in 1830 by Isambard Kingdom Brunel)

Figure 4.48 A selection of nuts – from left to right: flange nut, dome nut (for decorative purposes), wing nut (should only be hand tightened) square lock nut, nylon nut (the nylon insert prevents loosening), extension nut, castellated nut (used with a locking pin that goes through a hole in the screw shaft)

Remember! There is a multimedia version of this textbook that includes additional images and interactive features: www.automotivett.org

Figure 4.49 Twin lock nuts

exist for certain needs, such as wing nuts for finger adjustment and captive nuts for inaccessible areas.

Use of two nuts to prevent self-loosening In normal use, a nut-and-bolt joint holds together because the bolt is under a constant tensile stress called the preload. The preload pulls the nut threads against the bolt threads, and the nut face against the bearing surface, with a constant force, so that the nut cannot rotate without overcoming the friction between these surfaces. Extra preload can be created when two nuts are locked together.

Screws/bolts A screw, or bolt, is a type of fastener characterised by a helical ridge, known as an external thread or just thread, wrapped around a cylinder. Some screw threads are designed to mate with a complementary thread, known as an internal thread, often in the form of a nut or an object that has the internal thread formed into it. Other screw threads are designed to cut a helical groove in a softer material as the screw is inserted (self-tapping). The most common uses of screws are to hold objects together and to position objects.

Screw or bolt heads A screw and a bolt will always have a head, which is a specially formed section on one end of the thread that allows it to be turned, or driven. Common tools for driving screws include screwdrivers and wrenches. The head is usually larger than the body of the screw, which

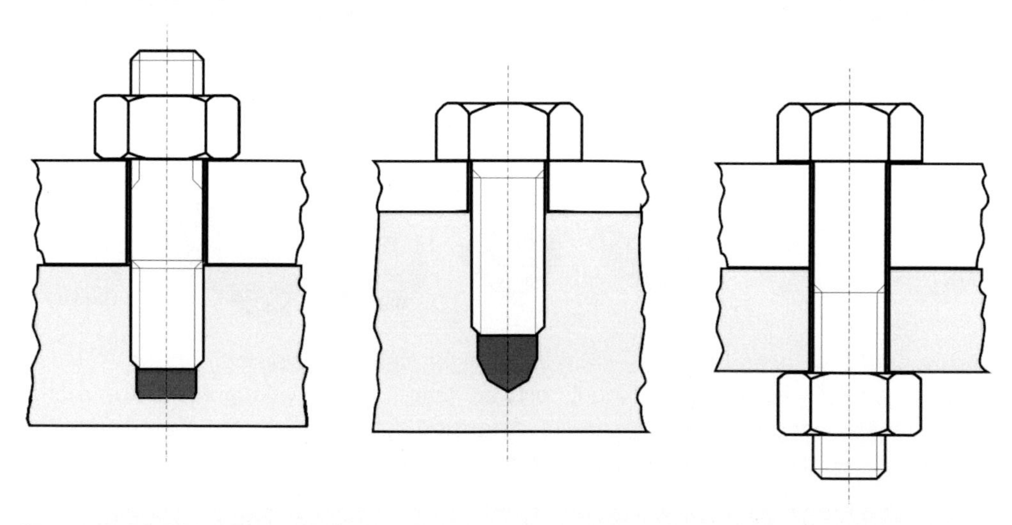

Figure 4.50 The difference between a stud, set screw (screw) and a bolt (from left to right)

Figure 4.51 Screw and a nut (with and Allen-type head)

keeps the screw from being driven deeper than the length of the screw and to provide a bearing surface. A selection of screws are shown here together with some common head shapes and the many different shapes for turning tools.

Figure 4.52 A selection of screws

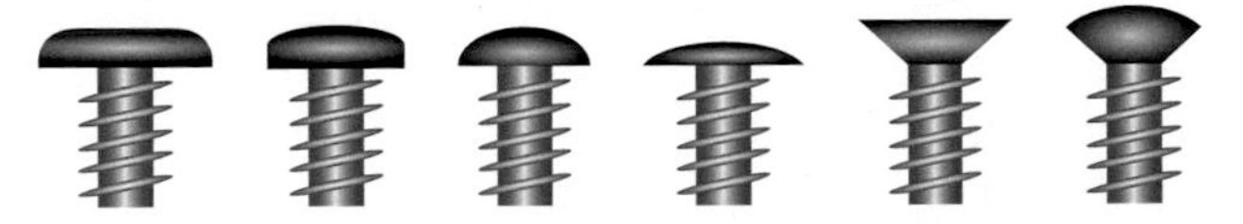

Figure 4.53 Screw heads from left to right: pan, dome (button), round, truss (mushroom), flat (countersunk), oval (raised head and countersunk)

Figure 4.54 Screw heads (from left to right): slot (flat), Phillips (ph), Pozidriv (pz), square (ext), square (int), hex, hex socket (Allen), security hex socket (pin-in-hex-socket), Torx (t & tx), security torx (tr), tri-wing, Torq-set, spanner head (snake-eye), triple square (xzn), polydrive, one-way, spline drive, double hex, Bristol, pentalobular

A typical nut and bolt is shown here. The cylindrical portion of the screw from the underside of the head to the tip is known as the shank; it may be fully or partially threaded (screw or bolt). The distance between each thread is called the 'pitch'. Screws and bolts are usually made of steel. Where great resistance to weather or corrosion is required, materials such as stainless steel, brass, titanium, bronze, silicon bronze or other specialist materials may be used. Alternatively, a surface coating is used to protect the fastener from corrosion. Selection criteria of the screw materials include: size, required strength, resistance to corrosion, joint material, cost and temperature.

Figure 4.55 Nuts and bolts with internal star lock washers

Tightening The majority of screws are tightened by clockwise rotation, which is termed a right-hand thread. Screws with left-hand threads are used in exceptional cases. For example, when the screw will be subject to counter clockwise torque (which would work to undo a right-hand thread), a left-hand-threaded screw would be an appropriate choice. The left side wheel nuts of many heavy vehicles use a left-hand thread.

Dimensions There are many systems for specifying the dimensions of screws, but in much of the world the ISO metric screw thread preferred series has displaced the many older systems. Other relatively common systems include the British Standard Whitworth, BA system (British Association), and the Unified Thread Standard.

A washer is a thin plate (usually disc-shaped) with a hole (most often in the middle) that is normally used to distribute the load of a threaded fastener, such as a screw or nut. Other uses include a spacer, spring, wear pad, preload indicating device, locking device and to reduce vibration (rubber washer). Washers usually have an outer diameter (OD) about twice the width of their inner diameter (ID). Automotive washers are usually metal or plastic. Washers are also important for preventing galvanic corrosion, particularly by insulating steel screws from aluminium surfaces.

Adhesives Another common method of securing threads is to use a locking compound such as 'Loctite'. This is in effect an adhesive which sticks the threads together. When the correct compound is applied with care, it is a very secure way of preventing important components from working loose.

Figure 4.56 Washers from left to right: spring (split), external star, wave, flat and flat

Figure 4.57 Loctite® Threadlocker (Source: © 2010 Henkel AG & Co. KGaA, Düsseldorf. All rights reserved)

Self-tapping screws As the name suggests these screws are designed to create their own thread as they are used. They are only suitable therefore for light duties and softer or thinner materials. A wide variety of sizes and head types are available; just three examples are show here. A captive spring nut can also be used with a self-tapping screw.

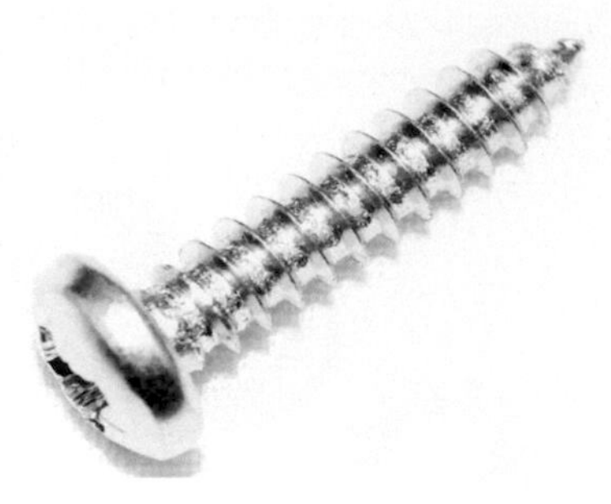

Figure 4.58 Phillips head self-tapping screw

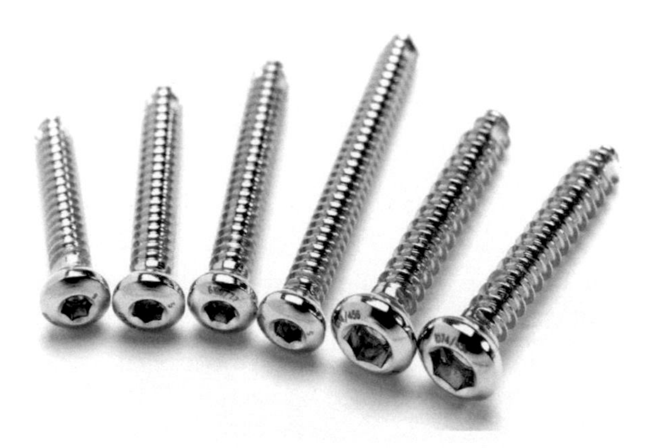

Figure 4.59 Allen head self-tapping screws

Figure 4.60 Combined slot and hex head self-tapping screw for heavier duty (may also be used with a spring-type captive nut)

Figure 4.61 Captive spring nut

Summary Nut screws washers and bolts . . . as the old newspaper headline goes! Seriously though, the types of fasteners or joining devices we have examined here are fundamental to automotive technology as well as all aspects of engineering. Make sure when using replacements that they are of equal quality as recommended by the vehicle manufacturer.

Look back over the previous section and write out a list of the key bullet points.

4.2 Workshop skills puzzles

4.2.1 Cryptograms

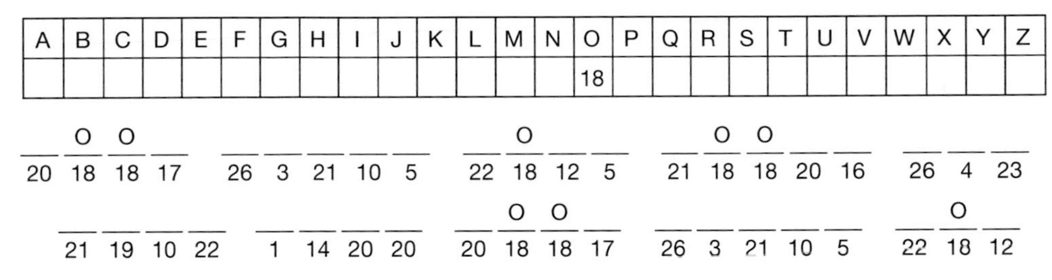

A	B	C	D	E	F	G	H	I	J	K	L	M	N	O	P	Q	R	S	T	U	V	W	X	Y	Z
														18											

_ O O _   _ _ _ _ _   _ O _ _   _ O O _ _   _ _ _
20 18 18 17   26 3 21 10 5   22 18 12 5   21 18 18 20 16   26 4 23

_ _ _ _   _ _ _ _   _ O O _   _ _ _ _ _   _ O _
21 19 10 22   1 14 20 20   20 18 18 17   26 3 21 10 5   22 18 12

Cryptogram 4.1 Hand tools rule:

4.2.2 Anagrams

Geraniums   Image Runs
Reaming Us   Mangier Us

Anagrams 4.1 Find out how much: what word did I use to get these anagrams?

4.2.3 Word search

M	E	Z	T	A	N	M	Y	L	E	O	S	T	M	E	
H	E	J	I	S	U	U	E	C	S	P	A	R	E	U	
I	G	L	A	Z	W	T	O	A	A	I	C	P	S	Z	
A	A	E	A	B	B	Q	D	N	S	R	L	X	H	G	
J	T	R	W	B	N	N	N	J	T	U	U	H	E	T	
E	L	H	I	C	I	E	F	C	J	U	R	C	E	J	ACCURACY
Q	O	M	I	C	R	O	M	E	T	E	R	I	C	L	BEARINGS
U	V	F	F	S	V	T	Q	A	H	E	O	T	N	A	CURRENT
I	O	M	O	J	X	Q	T	C	A	A	N	P	F	G	EQUIPMENT
P	A	K	Y	Y	S	V	I	W	B	E	M	W	G	E	HAMMER
M	X	I	G	H	P	J	A	S	R	D	K	M	G	L	MEASURING
E	F	S	D	S	G	N	I	R	A	E	B	J	E	Q	MICROMETER
N	W	R	E	N	C	H	U	Y	W	C	H	C	N	R	SPANNERS
T	W	S	P	A	B	C	R	P	Q	K	H	X	H	Y	VOLTAGE
B	N	W	B	N	C	R	P	P	B	G	L	C	L	S	WRENCH

Word search 4.1 Hand tools and measuring

4.2.4 Crosswords

Across

1 An instrument for measuring the make-up of a petrol engine exhaust (7,3,8)

5 A wrench driven by compressed air (3,7)

6 An electrical measuring tool with functions for volts, amps and ohms

10 A machining process that uses a rotating blade to cut through the surface of materials

13 Vehicle support equipment used to support a vehicle when wheels are removed (4,5)

15 Carborundum grit on a backing sheet (5,5)

Down

2 Prevents fluid leaks from a rotating shaft (3,4)

3 Diagnostic tool used to read fault codes and other data

4 Instrument for measuring electrical current

7 A simple bulb used for tracing circuit faults (4,4)

8 Device used to measure the exhaust from a diesel engine (5,5)

9 To lift something or a term used to describe a vehicle lift or ramp

10 Measuring device based on an accurate screw thread

4

16 Instrument for measuring electric potential

17 Used to tighten fasteners to a specified setting (6,6)

19 A mechanical or hydraulic tool for lifting a vehicle

20 An instrument that draws an image to represent changes in voltage over a set time

11 A material that is fitted between faces to provide a gas or liquid seal

12 A tool for applying a force to remove a gear from a shaft

14 A flashing light used to adjust ignition timing

18 An instrument used to indicate movement of a component on a clock display (4,5)

20 On board diagnostics (1,1,1)

Crossword 4.1 Hand tools, measuring and equipment

CHAPTER 5

Maintenance

5.1 Routine vehicle checks

Together with the multimedia resources, this section is ideal material for students working towards the IMI Awards unit EL11, the maintenance aspects of City and Guilds 3902 units and all other similar foundation or introduction level qualifications.

Automotive Technician Training: Entry Level 3. 978-0-415-72040-3.
© Tom Denton. Published by Taylor & Francis. All rights reserved.

After successful completion of this section you will be able to show you have achieved these objectives:

- Know vehicle systems and components that require routine checks.
- Know the information and equipment required for vehicle maintenance checks.
- Be able to safely and correctly carry out routine vehicle checks.

5.1.1 What you must know about vehicle checks

Vehicle systems and components that require routine maintenance: battery, engine oil, engine coolant, drive belts, fluid levels, wheels and tyres, lighting system, driver and passenger area; seating, seat belts, horn, instruments, warning lamps, door hinges, locks, mirrors, bodywork, paintwork, transmission; inspect for leakage

The tools and equipment required for vehicle maintenance: tyre tread gauge, inflator, pressure gauge, and car jack or trolley jack, and disposable cloths

Always wear suitable PPE and use tools and equipment safely and correctly when doing practical work on any vehicle

The maintenance requirements for vehicle systems:

- *check engine oil condition and level*
- *check engine oil filter condition and for leakage*
- *checking and top-up fluid levels; windscreen washer, battery, clutch and brake fluid*
- *checking and adjusting drive belts*
- *tyre condition, pressures and tread depth*
- *operation of vehicle lamps and indicators*
- *operation and condition of seat belts and seats*
- *operation of instruments, horn and warning lamps*
- *lubrication of door hinges and locks*
- *operation and condition of door mirrors*
- *condition of bodywork and paintwork*
- *checking and top-up transmission levels*

The information required for vehicle maintenance: vehicle make, model and VIN number, correct engine oil, specifications, engine coolant specifications, brake and clutch fluid specifications, specifications for new components or fluids, bulbs, and transmission

5.1.2 Main systems

No matter how we categorize them, all vehicle designs have similar major components and these operate in much the same way. The four main areas of a vehicle are the engine, electrical, chassis and transmission systems.

Engine: This area consists of the engine itself together with fuel, ignition, air supply and exhaust systems. In the engine, a fuel air mixture enters through an inlet manifold and is fired in each cylinder in turn. The resulting expanding gases push on pistons and connecting rods which are on cranks, just like a cyclists legs driving pedals, and this makes a crankshaft rotate. The pulses of power from each piston are smoothed out by a heavy flywheel. Power leaves the engine through the flywheel, which is fitted on the end of the crankshaft, and passes to the clutch. The spent gases leave via the exhaust system.

Figure 5.1 Ford Focus engine (Source: Ford Media)

Electrical: The electrical are covers many aspects such as lighting, wipers and instrumentation. A key components is the alternator which, driven by the engine, produces electricity to run the electrical systems and charge the battery. A starter motor takes energy from the battery to crank over and start the engine. Electrical components are controlled by a range of switches. Electronic systems use sensors to sense conditions and actuators to control a variety of things – in fact on modern vehicles, almost everything.

Chassis This area is made up of the braking, steering and suspension systems as well as the wheels and tyres. Hydraulic pressure is used to activate the brakes to slow down or stop the vehicle. Rotating discs are

Remember! There is a multimedia version of this textbook that includes additional images and interactive features: www.automotivett.org

Figure 5.2 A modern alternator (Source: Bosch Press)

gripped between pads of friction lining. The hand brake uses a mechanical linkage to operate parking brakes. Both front wheels are linked mechanically and must turn together to provide steering control. The most common method is to use a rack and pinion. The steering wheel is linked to the pinion and as this is turned it moves the rack to and fro, which in turn moves the wheels. Tyres also absorb some road shock and play a very important part in road holding. Most of the remaining shocks and vibrations are absorbed by

Figure 5.3 Disc brakes and part of the suspension system

springs in the drivers and passengers seats. The springs can be coil type and are used in conjunction with a damper to stop them oscillating (bouncing up and down too much).

Transmission In this area, the clutch is to allow the driver to disconnect drive from the engine and move the vehicle off from rest. The engine flywheel and clutch cover are bolted together so the clutch always rotates with the engine and when the clutch pedal is raised, drive is passed to the gearbox. A gearbox is needed because an engine produces power only when turning quite fast. The gearbox allows the driver to keep the engine at its best speed. When

Figure 5.4 Differential and final drive components

the gearbox is in neutral, power does not leave it. A final drive assembly and differential connect the drive to the wheels vial axles or driveshafts. The differential allows the driveshafts and hence the wheels to rotate at different speeds for when the vehicle is cornering.

Summary The layout of a vehicle, such as where the engine is fitted and which wheels are driven vary, as do body styles and shapes. However, the technologies used in the four main areas of a vehicle are similar no matter it is described. These are the:

- engine system
- electrical system
- chassis system
- transmission system.

These areas are covered in detail and make up the four main chapters of this book.

Create a mind map to illustrate the important features of a component or system in this section.

Look back over the previous section and write out a list of the key bullet points.

Now complete the multiple choice quiz associated with this topic/subject area.

5.1.3 Service sheets

Service sheets are used and records must be kept because they:

- Define the work to be carried out.
- Record the work carried out.
- Record the time spent.
- Record materials consumed.
- Allow invoices to be prepared.
- Record stock which may need replacing.
- Form evidence in the event of accident or customer complaint.

The following table is an example of a service sheet showing tasks carried out and at what service intervals. Please note once again that this list, while quite comprehensive, is not suitable for all vehicles, and the manufacturer's recommendations must always be followed. Some of the tasks are only appropriate to certain types of vehicle. The following table also lists the work in a recommended order, including the use of a lift.

Driving vehicle into workshop	
Instrument gauges, warning/control lights and horn	Check operation
Washers, wipers	Check operation/adjust, if necessary
Inside vehicle	
Exterior and respective control lights; instrument cluster illumination	Check operation/condition
Service interval indicator	Reset after every oil change if applicable
Handbrake	Check operation/adjust, if necessary
Seat belts, buckles and stalks	Check operation/condition
Pollen filter	Renew
Warning vest	Check availability – if applicable
First aid kit	Check availability and expiry date – if applicable
Warning triangle	Check availability – if applicable
Outside Vehicle	
Hood latch/safety catch & hinges +	Check operation/grease
Road (MOT) test	Check regarding next road (MOT) test due date – if applicable
Emission test	Check regarding next emission test due – if applicable
Under bonnet/hood	
Wiring, pipes, hoses, oil and fuel feed lines	Check for routing, damage, chafing and leaks
Under bonnet/hood	
Engine, vacuum pump, heater and radiator	Check for damage and leaks
Coolant	Check anti-freeze concentration: °C

Coolant expansion tank and washer reservoirs	Check/top up fluid levels as necessary – in case of abnormal fluid loss, a separate order is required to investigate and rectify
Power steering fluid	Check/top up fluid levels as necessary – in case of abnormal fluid loss, a separate order is required to investigate and rectify
Battery terminals	Clean, if necessary/grease
Battery	Visual check for leaks -in case of abnormal fluid loss, a separate order is required to investigate and rectify
Fuel filter	Drain water, if not renewed – diesel models (with drain facility)
Headlamp alignment	Check – adjust alignment, if necessary
Brake fluid	Check/top up fluid levels as necessary -in case of abnormal fluid loss, a separate order is required to investigate and rectify
Under vehicle	
Engine	Drain oil and renew oil filter
Steering, suspension linkages, ball joints, sideshaft joints, gaiters	Check for damage, wear, security and rubber deterioration
Engine, transmission	Check for damage and leaks
Pipes, hoses, wiring, oil and fuel feed lines, exhaust	Check for routing, damage, chafing and leaks
Underbody	Check condition of PVC coating
Tyres	Check wear and condition, especially at tyre wall, note tread depth: RF mm, LF mm, LR mm, RR mm, Spare mm
Brake system	With wheels off, check brake pads, discs, linings for wear and check brake cylinders for condition: check rubber components for deterioration.

Use a library or the interactive web search tools to examine the subject in this section in more detail.

5.1.4 Effects of incorrect adjustments

The following table lists a selection of possible incorrect adjustments, together with their effects on the operation of the vehicle. This is intended to be an exercise to help you see why correct adjustments are so important. Not so you know how to do it wrong! You must also be able to make a record and tell a customer the effects, if you are unable to make the correct adjustments. This could be due to some parts being worn so that adjustment is not possible.

Remember though, anyone can mess with a vehicle and get it wrong. As a professional you will get it right, the customer and your company will be happy and it will affect your pay rates in years to come.

One of the problems, which can arise after a vehicle has been serviced, is when the customer expected a certain task to be completed – but it was not. For

Incorrect adjustment	Possible effects						
Brake	Excessive pedal and lever travel	Reduced braking efficiency	Un-balanced braking	Over-heating	Skidding and a serious accident		
Drive belts	Over-heating	Battery recharge rate slow.	Power steering problems	AC not operating			
Fuel system	Poor starting or non-start	Lack of power or hesitation	Uneven running & stalling	Popping back or backfire	Running on or detonation	Heavy fuel usage	Fuel leaks and smells
Ignition	Poor starting or non-start	Lack of power	Hesitation	Exhaust emission	Running on		
Plug gaps	Poor starting or non-start	Lack of power	Hesitation	Uneven running	Misfire	Exhaust emissions	
Steering system	Abnormal or uneven tyre wear	Heavy steering	Pulling to one side	Poor self-centring	Wandering	Steering wheel alignment	Excessive free play
Tyre pressures	Abnormal tyre wear	Heavy steering	Uneven braking	Heavy fuel usage	Tyre life time is reduced		
Valve clearances	Lack of power	Uneven running	Misfire	Excessive fuel usage	Exhaust emissions.	Noise from valves or camshaft	

example on a basic, interim or even in some cases a full service, little or no work is carried out on the ignition system. This will not therefore rectify a misfire. It is important that the customer is aware of what will be done, as well as what was done to their vehicle. And, if during a service, you notice a fault – report it.

Use a library or the interactive web search tools to examine the subject in this section in more detail.

5.1.5 Information sources

The main sources of information are:

- technical manuals
- technical bulletins
- servicing schedules
- job card instructions
- inspection records
- checklists.

All main manufacturers now have online access to this type of information. It is essential that proper documentation is used and that records are kept of the work carried out. For example:

- job cards
- stores and parts records
- manufacturers' warranty systems.

These are needed to ensure the customer's bill is accurate and also so that information is kept on file in case future work is required or warranty claims are made.

Qty.	Part Number and/or Description	Unit Cost	Amount
	TOTAL PARTS		
Sublet repairs:			
	TOTAL SUBLET REPAIR		

Date IN | Repair Order #
Date OUT | Written By

Name (or training aid #) | Home Phone Number | Business Phone Number
Address
City | ST | Zip | License Plate Reg. Number | Vehicle Colour/Description
Year | Make | Model | Vehicle Odometer Mileage | Technician Name or Number
V.I.N. →

SERVICES REQUESTED and DESCRIPTION OF WORK	AMOUNT
concern:	
cause:	
correction:	

ESTIMATED COSTS
Parts | Labour | Total
Authorized by | In Person ☐ | By Phone ☐
Date | Time | Contacted by | Phone #

REVISED ESTIMATE or ADDITIONAL WORK
Parts | Labour | Total
Authorized by | In Person ☐ | By Phone ☐
Date | Time | Contacted by | Phone #
(use the back for calculations and additional information)

TOTAL LABOUR (calculated from the flat rate manual)
TOTAL PARTS
TOTAL REPAIR
Simulated TAX
Suggested TOTAL DONATION

I hereby authorize the requested diagnostic, repair and/or service operation(s) to be performed by students in an educational setting pertaining to the current curriculum. The vehicle may be driven or otherwise used to effectuate these operations. It is understood that this is considered a training exercise with NO implied warranty and the institution will not be held responsible for loss or damage to the vehicle or contents due to fire, theft, or any other cause beyond control. An express "mechanics lien" is acknowledged on this vehicle in order to secure the amount of repairs thereto.

signature

save old parts? ☐ yes

Figure 5.5 Example job sheet /repair order

Results of any tests carried out will be recorded in a number of different ways. The actual method will depend on what test equipment was used. Some equipment will produce a printout for example. However, results of all other tests should be recorded on the job card. In some cases this may be done electronically but it is the same principle. Remember: always make sure that the records are clear and easy to understand.

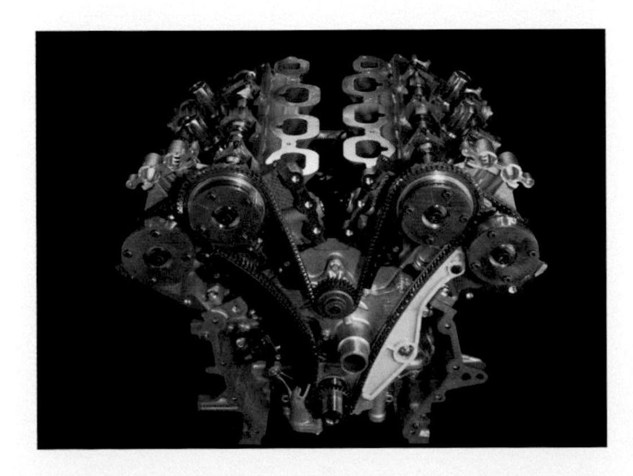

Figure 5.6 V6 Mustang engine – data is essential when working on complex systems

Look back over the previous section and write out a list of the key bullet points.

Now complete the multiple choice quiz associated with this topic/ subject area.

Now complete the multiple choice quiz associated with this **unit**.

5.2.0 Routine vehicle maintenance processes and procedures

Together with the multimedia resources, this section is ideal material for students working towards the IMI Awards unit L108, the maintenance aspects of City and Guilds 3902 units and all other similar foundation or introduction level qualifications.

After successful completion of this section you will be able to show you have achieved these objectives:

- Be able to work safely.
- Know vehicle components and systems that require routine maintenance.
- Know routine maintenance requirements for vehicle systems and components.
- Be able to carry out routine vehicle maintenance e.g. an interim service.

5.2.1 What you must know about vehicle inspection

Components that require routine inspection:

- *tyres – wear and condition*
- *wheels – damage, buckling*
- *brakes – wear, adjustment, fluid leaks, fluid level, corrosion of pipes, condition of hoses*
- *steering and suspension – security of components, wear of joints, suspension damper*
- *electrical – battery, alternator, warning lamps, front and rear wipers, horn*
- *lighting – function of side and rear lamps, number plate lamp, headlamps, dip and main beam control, boot lamp (on and off), interior lamps, indicators, hazard lamps, front and rear fog lamps*
- *engine compartment – washer fluid, brake fluid level, coolant leaks and level, oil leaks and level, bonnet release, battery, drive belts*
- *transmission – clutch operation and adjustment, drive shafts, joints, rubber boots, fluid leaks*
- *vehicle exterior – bodywork, paintwork, trim, doors and door locks, wing mirror condition*
- *vehicle interior – seats (condition & adjustment), seat belts, driver controls, warning lamps, wing mirror operation*

Technical data and information: specifications and data, manufacturer's inspection requirements, manuals, inspection check lists, torque settings, lubricant grades, quantities, tyre pressures, and legal requirements

Always wear suitable PPE and use tools and equipment safely and correctly when doing practical work on any vehicle

Replacement and replenishment of fluids: brake fluid, power steering fluid, battery electrolyte, washer fluid, coolant, oil and filter change and level, and transmission fluids

Adjustments and lubrication of components that require routine maintenance: doors and door locks, bonnet catch and hinges, headlamps, tyre pressures, belts – alternator and water pump

Procedures: following inspection schedule, not assuming component and system is satisfactory, systematic visual inspection of components, systematic functional checks on components and system operation, use of aural assessment during system operation, visual signs and indicators of wear, maladjustment and corrosion, using measurement tools, comparison of measurements with specifications, using levers and bars and recording possible concerns

5.2.2 Introduction

It is important to carry out regular servicing and inspections of vehicles for a number of reasons:

- Ensure the vehicle stays in a safe condition.
- Keep the vehicle operating within tolerances specified by the manufacturer and regulations.
- Ensure the vehicle is reliable and reduce down time.
- Maintain efficiency.
- Extend components and the vehicles life.
- Reduce running costs.
- Keep the vehicle looking good and limit damage from corrosion.

Figure 5.7 Servicing the brakes

In order to carry out servicing and inspections you should understand how the vehicle systems operate. It is also important to keep suitable records; this is often known as the vehicles' service history. Services and inspections of vehicles vary a little from one manufacturer to another. Servicing data and servicing requirement books are available as well as the original manufacturer's information. This type of data should always be read carefully so as to ensure that all the required tasks are completed. The following table lists some important words and phrases relating the servicing and inspection.

First service	This service is becoming less common but some manufacturers like the vehicles to be returned to the dealers after about a thousand miles or so. This is so that certain parts can be checked for safe operation and in some cases oil is changed

Distance-based services	Ten or twelve or twenty thousand mile intervals are common distances but manufacturers vary their recommendations. Most have specific requirements at set distances
Time-based services	For most light vehicles, distance-based services are best. Some vehicles though run for long periods of time but do not cover great distances. In this case the servicing is carried out at set time intervals. This could be every six months, six weeks or after a set number of hours run.
Inspection	The MOT test which must be carried out each year after a light vehicle is older than three years is a good example of an inspection. However an inspection can be carried out at any time and should form part of most services.
Records	A vital part of a service. To ensure all aspects are covered and to keep information available for future use.
Customer contracts	When you make an offer to do a service, and the customer accepts the terms and agrees to pay, you have made a contract. Remember that this is legally enforceable by both parties.

Figure 5.8 Bodywork protection in use during repairs

Clearly, it is important to keep a customer's vehicle in a clean condition. To do this there are a number of methods as outlined here:

- Seat covers to keep the seats clean.
- Floor mats to protect the carpets.
- Steering wheel covers to keep greasy hand prints off the wheel.
- Wing covers to keep the paintwork clean and to prevent damage.

Look back over the previous section and write out a list of the key bullet points.

5.2.3 Maintenance and inspections

The purpose of routine maintenance is simple; it is to keep the vehicle in a good working order and in a safe condition. Manufacturers specify intervals and set tasks that should be carried out at these times. It is usually a condition of the warranty that a vehicle should be serviced according to manufacturers' needs. The main purpose of regular inspection therefore, is to check for the following:

- malfunction of systems and components
- damage and corrosion to structural and support regions
- leaks
- water ingress
- component and system wear and security.

Inspections are usually:

- aural – listening for problems
- visual – looking for problems
- functional – checking that things work!

The main types of inspection, in addition to what is carried out when servicing, are:

- pre-work

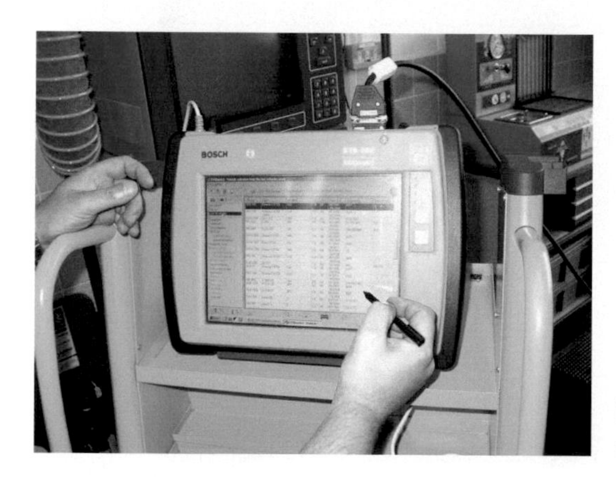

Figure 5.9 Checking data and setting up test equipment

- post-work
- pre-delivery inspection (PDI)
- used vehicle inspection
- special inspection (maybe after an accident for example).

A pre-work inspection is used to find out what work needs to be carried out on a vehicle. Post-work inspections are done to make sure the repairs have been carried out correctly and then no other faults have been introduced.

A PDI is carried out on all new vehicles to check certain safety items and to, for example, remove any transport packaging such as suspension locks or similar. A used vehicle inspection is done to determine the safety and saleability of a vehicle as well as checking that everything works. After gaining experience you may be asked to carry out an inspection of a vehicle after an accident to check the brakes condition for example.

In all cases, a recommended checklist should be used and careful records of your findings should be kept. Working to timescales, or reporting to a supervisor that timescales cannot be met, is essential:

1 First, when a customer books a car in for work to be done they expect it to be ready at the agreed time. Clearly if this deadline can't be met the customer needs to be informed.
2 Second, in order to make the running of a workshop efficient and profitable, a technician will have jobs allocated that will take a certain amount of time to complete. If for any reason this time can't be met then action will need to be taken by the workshop manager or supervisor.

The three main **regulations** that cover the repair and service of motor vehicles in the UK are:

1 Road Traffic Act – this covers things like road signage and insurance requirements. It also covers issues relating to vehicle safety. For example, if a car suspension was modified it may become unsafe and not conform to the law.
2 VOSA regulations – the main one of these being the annual MOT test requirements. VOSA stands for Vehicle and Operator Services Agency.
3 Highway Code – which all drivers must follow and forms part of the driving test.

Similar regulations are in place in other countries. The regulations are designed to improve safety.

Some of the main vehicle systems relating to safety are listed in the table, together with examples of the requirements. Note though that these are just examples and that specific data must be studied relating to specific vehicles.

Brakes	The foot brake must produce 50% of the vehicle weight braking force and the parking brake 16% (this assumes a modern dual line braking system). The brakes must work evenly and show no signs of leaks.
Exhaust	Should not leak, which could allow fumes into the vehicle, and it should not be noisy!
Horn	It should be noisy!
Lights	All lights should work and the headlights must be correctly adjusted.
Number plates	Only the correct style and size must be fitted. The numbers and letters should also be correctly spaced and not altered (DAN 15H is right, DANISH is wrong)!

Remember! There is a multimedia version of this textbook that includes additional images and interactive features: www.automotivett.org

Seat belts	All belts must be in good condition and work correctly.
Speedometer	Should be accurate and illuminate when dark.
Steering	All components must be secure and serviceable.
Tyres	Correct tread depth is just one example.

After successful completion of this section you will be able to show you have achieved these objectives:

- Know commonly used valeting tools and equipment and how they are used correctly and safely.
- Know the commonly used cleaning materials and how they are used correctly and safely.
- Be able to demonstrate the correct procedures for valeting motor vehicles safely and effectively.

Windscreens and other glass	You should be able to see right through this one! No cracks allowed in the screen within the driver's vision.

Use a library or the interactive web search tools to examine the subject in this section in more detail.

Look back over the previous section and write out a list of the key bullet points.

5.3.0 Basic vehicle valeting

Together with the multimedia resources, this section is ideal material for students working towards the IMI Awards unit EL12, the City and Guilds 3902–010 unit and all other similar foundation or introduction level qualifications.

5.3.1 What you must know about valeting

Always wear suitable PPE and use tools and equipment safely and correctly when doing practical work on any vehicle

The correct sequence and procedure for valeting a vehicle's exterior:

1. *wetting vehicle with water hose including the wheels and arches*
2. *washing vehicle from top to bottom using shampoo and sponges including door entries and boot entry*
3. *hosing vehicle to remove shampoo*
4. *drying paintwork and glass using chamois leather*
5. *removal of tar from paintwork using appropriate cleaner*
6. *cleaning alloy wheels using appropriate cleaner*
7. *cleaning glass surfaces*
8. *cleaning tyres*
9. *cleaning chrome surfaces*
10. *applying polish to paintwork avoiding direct sunshine*
11. *buffing paintwork after polishing*

The correct sequence and procedure for valeting a vehicle's interior:

1. *cleaning and brushing seats and upholstery*
2. *cleaning glass surfaces*
3. *cleaning plastic and dashboard surfaces*
4. *removal of car mats*
5. *vacuuming seats and carpets*
6. *reinstate car mats and property*
7. *clean boot area*

The materials used for valeting a vehicle's exterior: shampoo, polish, tyre blackener, glass cleaner, tar remover, chrome cleaner, and alloy wheel cleaner

The materials used for valeting a vehicle's interior: upholstery cleaner, shampoo, glass cleaner, dashboard cleaner and carpet shampoo

Tasks and precautions: manufacturers' recommendations, instructions, appropriate materials for surface, avoiding contamination or splashing of other surfaces and avoiding the use of contaminated cloths

Tools and equipment for valeting: water hose, power hose, cleaning brushes for paintwork, wheel brushes or scrubbers, sponges and buckets, chamois leather, polishing cloth, cleaning cloth, upholstery brush and vacuum

Tasks to prepare and use valeting tools safely and correctly: electrical safety of power hoses and vacuum cleaners, ensuring sponges and cleaning cloths are free of grit and dirt, soaking and squeezing chamois leather for drying surfaces

5.3.2 Overview, equipment and safety

Introduction When valeting a vehicle it is worth considering the reasons for such an operation. Why should a vehicle be presented to such a high standard? I suggest that there are three main reasons:

1. It enhances the value of the vehicle.
2. A customers first impressions are very important.
3. It simply looks good.

A vehicle will be valeted for a number of reasons:

- A request from a customer.
- Preparation for sale.
- Part of a vehicle service as an incentive to the customer.

A wide range of cleaning products is available to assist with the valeting process.

Figure 5.10 Valeting products (Source: Autoglym)

Safety and PPE The products for vehicle valeting are chosen to do a job quickly and to a good standard but this must never be at the expense of safety. Let's start with the obvious safety recommendations and remember that safety precautions are for your benefit:

- Only use a product for its intended purpose.
- Store cleaning materials safely and away from children.
- Use appropriate PPE.
- Always follow manufacturers' instructions.

Figure 5.11 Safety glasses

Data sheets Detailed data sheets are available from the manufacturers on request. This is a requirement of the COSHH regulations.

Because of the job they have to do valeting products are made from strong chemicals, take extreme care.

The data sheet show here is available from the Autoglym website.

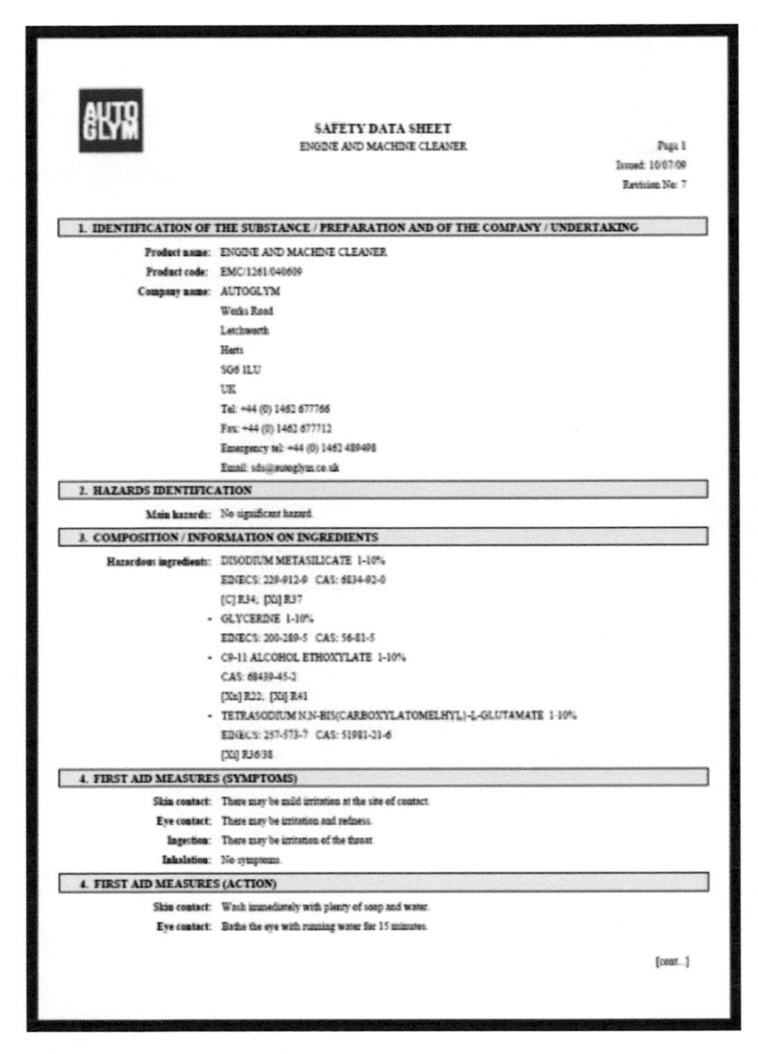

AUTO GLYM

SAFETY DATA SHEET

ENGINE AND MACHINE CLEANER

Page 1
Issued: 10/07/09
Revision No: 7

1. IDENTIFICATION OF THE SUBSTANCE / PREPARATION AND OF THE COMPANY / UNDERTAKING

Product name: ENGINE AND MACHINE CLEANER
Product code: EMC/1261/040609
Company name: AUTOGLYM
Works Road
Letchworth
Herts
SG6 1LU
UK
Tel: +44 (0) 1462 677766
Fax: +44 (0) 1462 677712
Emergency tel: +44 (0) 1462 489498
Email: sds@autoglym.co.uk

2. HAZARDS IDENTIFICATION

Main hazards: No significant hazard.

3. COMPOSITION / INFORMATION ON INGREDIENTS

Hazardous ingredients: DISODIUM METASILICATE 1-10%
EINECS: 229-912-9 CAS: 6834-92-0
[C] R34; [Xi] R37
- GLYCERINE 1-10%
EINECS: 200-289-5 CAS: 56-81-5
- C9-11 ALCOHOL ETHOXYLATE 1-10%
CAS: 68439-45-2
[Xn] R22; [Xi] R41
- TETRASODIUM N,N-BIS(CARBOXYLATOMELHYL)-L-GLUTAMATE 1-10%
EINECS: 257-573-7 CAS: 51981-21-6
[Xi] R36/38

4. FIRST AID MEASURES (SYMPTOMS)

Skin contact: There may be mild irritation at the site of contact.
Eye contact: There may be irritation and redness.
Ingestion: There may be irritation of the throat.
Inhalation: No symptoms.

4. FIRST AID MEASURES (ACTION)

Skin contact: Wash immediately with plenty of soap and water.
Eye contact: Bathe the eye with running water for 15 minutes.

[cont...]

Figure 5.12 Data sheet example (Source: Autoglym)

Pressure washers Pressure washers can increase the water pressure from about 4 bar up to 100 bar. This can actually save water, as the washing process is much faster. Steam cleaners are similar except that they use high pressure and very high temperature water. The water is pressurised by an electric pump and heated by burning a fuel.

Steam cleaner safety The potential for accidents is significant if safety procedures are not followed. The steam cleaner has the following safety risks:

- Very hot water – risk of burns or scalding.
- High-pressure water – damage to the vehicle or eyes and ears for example.
- Hot machinery – the heating coils get very hot and will burn.
- Naked flames – burning paraffin is often used to create the hot water.

- Strong detergents – skin and eye damage.
- High voltage electrical supply – serious risk as it is in the vicinity of water.
- Metal lance – gets very hot but can also damage the vehicle paintwork.

Figure 5.13 Pressure washer (Source: Karcher)

Figure 5.14 Steam cleaner (Source: Karcher)

Polishing tools These are useful for exterior cleaning. They can save a lot of time when polishing or buffing up. An important precaution, however, is to take extra care that you do not damage the paintwork.

Interior cleaning equipment The main piece of equipment for interior cleaning is the wet vacuum cleaner. Water and suitable cleaning fluid is forced into the upholstery under pressure. A strong vacuum is then used to suck out the water and dirt. The water is collected in the machine for later disposal. Carpet cleaners used in the home are very similar. A normal but heavy-duty vacuum cleaner is also an essential interior cleaning tool.

Figure 5.15 Polisher (Source: Craftsman Tools)

Figure 5.16 Wet or Dry Vacuum (Source: Karcher)

Summary Remember, valeting:

1. Enhances the value of a vehicle.
2. Makes a good first impression.
3. Ensures that the car looks its best!

Always follow manufacturers' procedures when using cleaning solutions and equipment.

5.3.3 Exterior cleaning

Cleaning the wheels The first step in washing a vehicle will often be to clean the wheels, tyres and wheel arches. If brake dust is allowed to remain on the wheel it can permanently bond with and etch the surface.

Follow the directions provided by the manufacturer. The wheel cleaner will often be sprayed on and worked in with a brush. Use a shampoo solution to remove the wheel cleaner. Make sure you do not allow the wheel cleaning solution to dry on the wheel and tyre surface. Some cleaning solutions will etch the wheel, if allowed to dry without rinsing. Thoroughly spray with water to rinse of all the cleaner to complete the job.

Washing the paint Washing the vehicle by hand with a bodywork shampoo is the best method. Do not use household detergents because they are made to remove grease. They will also remove the polish on the paint surface and accelerate the oxidation process. A pressure washer is an ideal tool to assist with washing.

Washing process To wash the vehicle:

1. Make sure that the paint surface is cool to the touch.
2. Wash from the top down.
3. Use either a natural fibre mitt or a shampoo sponge.
4. Follow the instructions regarding the amount of shampoo to use.
5. Wash the vehicle in small sections with frequent rinsing to prevent the water and the contaminants that you are removing from drying on the surface of the paint.
6. Rinse your mitt or sponge frequently as you progress down the sides of the vehicle, since there is more dirt and contaminants closer to the ground.
7. Finish with a complete rinse of the entire vehicle.

Figure 5.17 Start at the top . . .

Figure 5.18 . . . and work down the sides

Figure 5.19 Rinsing

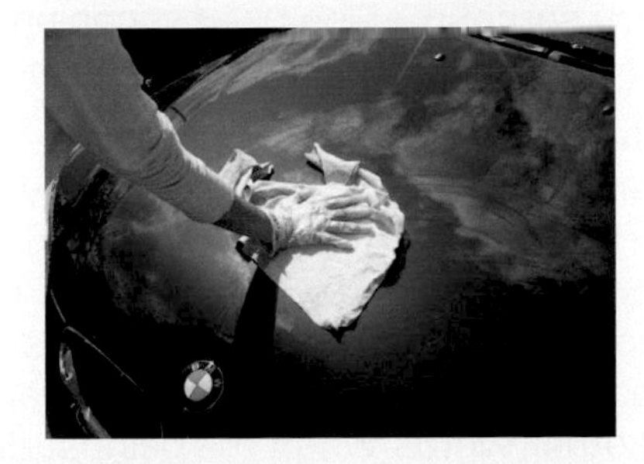

Figure 5.20 Drying with a chamois leather

Drying There are five steps in the drying process to get the best results:

1. Remove the largest volume of the water from the entire vehicle with a chamois.
2. Blow out all the channels where water can accumulate.
3. Further dry all surfaces, including windows and wheels.
4. Open all the doors, the engine compartment and the boot to eliminate any remaining run off and water tracks.
5. Wind the windows down a little to clean any dirt from the seal area at the top and sides of the glass.

Inspection When the drying is complete you are ready to inspect the exterior surface. Depending on the condition of the paint, there are several different products that may be needed to produce an ideal finish. Examples of these include:

- wax and grease strippers
- solvents and cleaners
- tar and road oil removers
- 'clay' – to remove paint overspray and other imbedded contaminants
- abrasives and glazes.

Tar and road film removers These products are harsh on the paint surface and care should be taken. They should not remain on the surface over a few seconds. These products may soften the paint and cause discoloration.

Figure 5.21 Tar remover

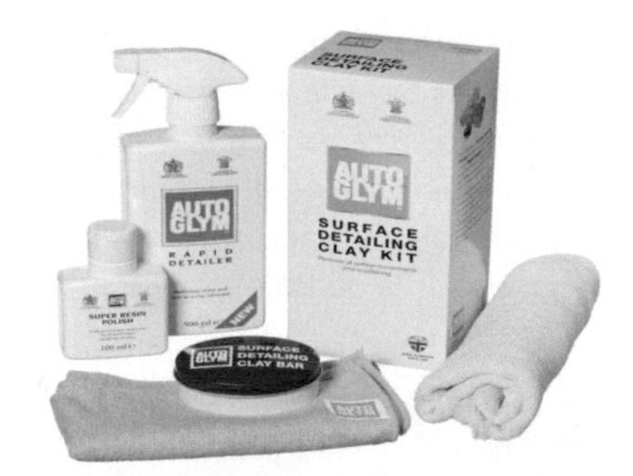

Figure 5.22 Clay kit (Source: Autoglym)

Clay products These products are used to remove paint overspray from the paint surface. They should be used with caution, as they are designed to lift contaminants and carry them away from the paint surface in the clay material. These products can often remove overspray without resorting to loss of the paint surface through use of compounds that contain abrasives. The clays may sometimes be used in conjunction with soapy water as a wetting agent for the gentle rubbing of the paint surface with the clay.

Abrasives compounds Caution – these products remove paint from the vehicle's surface. There is no way to put the paint back on when you go too far! The range of abrasive grit can be thought of as roughly comparable to wet and dry rubbing down paper. The types of paint damage that will require this product include heavy oxidation, serious water spotting, paint overspray that has bonded and several types of air pollution that contain acid components. In other words when the paint surface itself has been damaged. The normal

Figure 5.23 Application to a small scratch

Figure 5.24 The end result

procedure is to cover a small area with a back and forth motion. The product should not be allowed to dry and should be removed when it becomes cloudy or tacky. Take care!

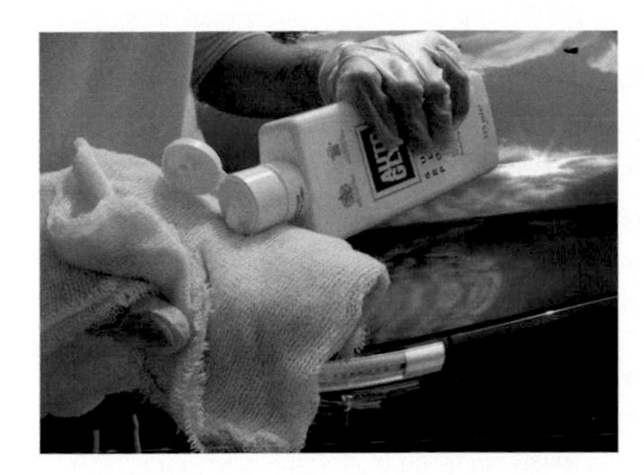

Figure 5.25 Apply polish to the cloth

Polymers and waxes There are two main products for protecting automotive paint finishes:

- polymers – synthetic, manmade substances
- waxes – naturally occurring substances found in trees (and bees!).

There is a much discussion regarding the merits of each product. The cost of most polymers will be higher than their counterpart waxes but polymers will last up to four times longer. This durability factor is a significant consideration. There is an edge to the polymers when comparing ease of use; they go on easier and come off faster. An additional benefit is the clarity and refractive consistency that the polymers produce on the finish surface. When comparing depth of gloss, however, high-quality waxes have an edge.

Finishing All finishing products should be applied in a front to back motion with a cotton cloth. The drying time can range from minutes to days, to achieve maximum bonding for some polymers. I suggest a product that dries in a few minutes will be most appropriate for professional valeting. All polymer and wax products should ideally be hand finished. However use of an orbital buffer with a high-quality pad can reduce the effort of removal of the finish product. In all cases, the final step in removal should be by hand.

Glass cleaning Products, which contain ammonia, do a good job of removing the dirt and film that is found on vehicle glass. One of the main problems with cleaning the inside surfaces is the build-up of contaminants. These come from the decomposition of the vinyl and plastic components. This decomposition and transfer to the surface of the glass is caused by the ultra violet action of the sun.

It is recommended that two cotton cloths are used to clean and then dry the surface. Spray the cleaner on one cloth to prevent the cleaner from getting on the surface of the dashboard as you wipe the glass. Use the second towel to

Figure 5.26 Clean the glass and polish with a dry cloth

Figure 5.27 Before washing the engine check carefully for areas that may be susceptible to water damage

dry the surface. Clean the outside first as this makes it easier to see where there are any streaks when you do the interior. Lower the door windows to get both sides of the glass, which is recessed, into the door.

Cleaning the engine First remove the debris that you will find in the channels of the body, bonnet and the grill openings. The best way to accomplish this job is with an air line. Next cover all electrical connections such as sensors, distributor and spark plug openings with either plastic bags or film bags. Use tape to seal the plastic surrounding these connections to prevent water from reaching them if necessary.

Completely wet down the wings, grill, top and bottom of the bonnet and the entire engine compartment. A pressure washer is ideal, but not essential. It is important to wet down all painted surfaces that surround the engine compartment, because the degreaser solution that you will use for cleaning may strip the protective coatings on these surfaces.

Note that if the engine is to be cleaned, it will normally be done before the rest of the vehicle.

Engine degreaser A pressure tank sprayer or a spray bottle for application of this product is best. Engine cleaners work most effectively when all of the surfaces to be cleaned have received a thorough soaking. The product should be allowed to stay on the engine components for a few minutes but check the instructions on the container. Use of a brush on the engine and other surfaces may help to remove heavy oil deposits if you are not using a pressure washer.

To remove the engine cleaner, completely soak down the entire compartment and the surrounding surfaces. If you use a pressure washer be careful not to get the nozzle too close to any electrical connectors. When the compartment has dried, spray on a rubber and plastic conditioner.

Figure 5.28 Engine cleaner

Rubber and plastic There are two causes of damage to exterior rubber and plastic surfaces:

- UV light
- ozone.

Most products that are designed to condition plastic and rubber have silicone as an ingredient. There is however some controversy regarding silicone as a component in both non-porous and porous surface conditioning. There are some people that believe that it will dissolve certain components in the rubber and cause cracking. It looks good when you first apply it though!

New car de-waxing When a new vehicle is delivered from the factory it is often protected with a type of wax or lacquer. This must be removed prior to the customer taking delivery. There are two main types of protection often referred to as:

- hard wax
- soft wax.

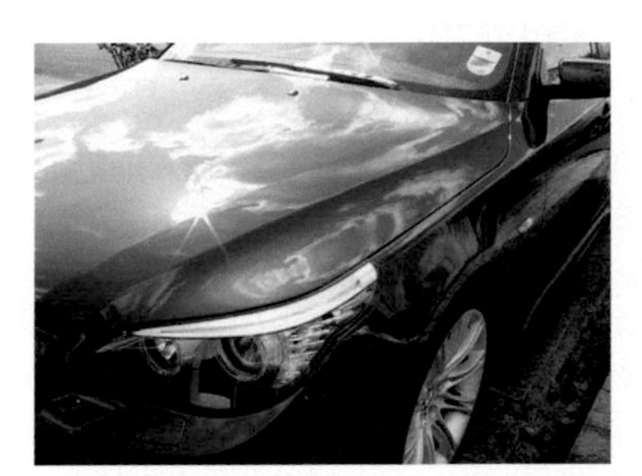

An appropriate solvent is used to soften the wax. The vehicle is then washed in the normal manner most often with a pressure washer or steam cleaner.

Figure 5.29 The finished job

Summary Always follow manufacturers' instructions relating to cleaning products and equipment. Finally, remember to give the vehicle a last look over to ensure there are no small marks or smears, because you can be sure that these would be the first thing your customer will see!

5.3.4 Interior cleaning

Vinyl and leather After vacuuming use a cloth and a spray bottle for the application of an interior cleaner. Be careful to test a small area of the surface before you begin to make sure that the product will not produce any damage or significant colour removal.

Following cleaning, the conditioning step will be essentially the same process. In most cases the manufacturers will recommend buffing the conditioning

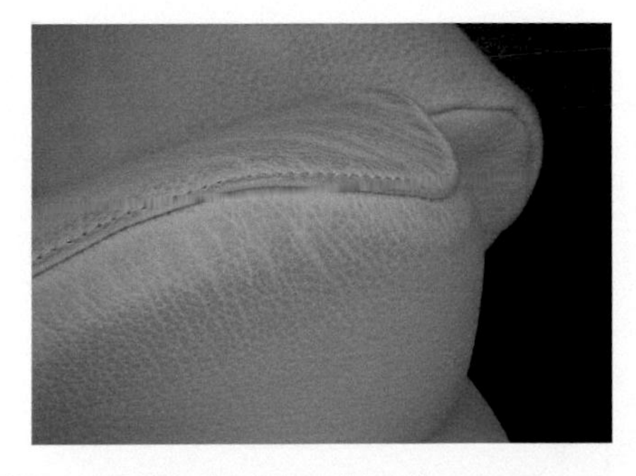

Figure 5.30 A stained leather seat

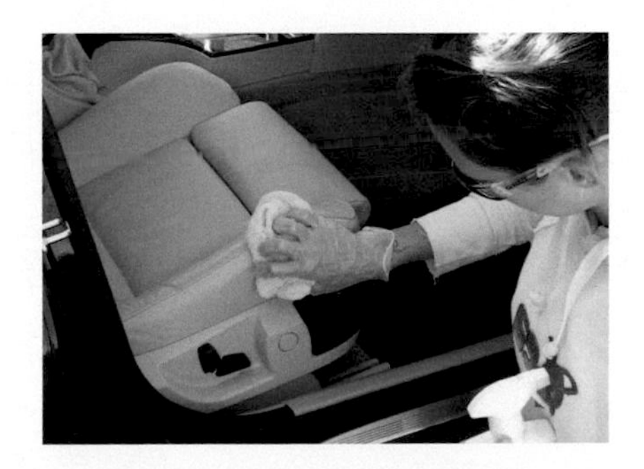

Figure 5.31 Cleaning a leather seat with special cleaner

Figure 5.32 Apply leather care product to a cloth and then 'feed' the leather after cleaning

Figure 5.33 Use a brush attachment for the dashboard area

product after it has had a few minutes to absorb into the surface. Regular cleaning and conditioning of vinyl and leather will extend the life of the product by a significant degree. It is particularly important to 'feed' leather regularly.

Carpet and upholstery products Care should be taken when using products for cleaning carpets and upholstery, to ensure that the product will not cause any unwanted result such as colour removal. The surface should be thoroughly vacuumed both before and after use of a cleaner. The benefit of through re-vacuuming is that it will remove the contaminants that the cleaner has lifted from the surface. Most professional valeters will use a wet vacuum with a detergent in the water. The fluid is sprayed in to the upholstery under pressure and then vacuumed out.

Air fresheners There is a wide range of scents available in deodorants/air fresheners. An aerosol spray, directly on the surfaces that will absorb the product is a good idea. These surfaces include the carpet, floor mats and the cloth upholstery (not vinyl and leather). Don't select products that have an overpowering aroma!

Summary As usual, always follow manufacturers' instructions relating to cleaning products and equipment. Finally, remember to fit seat covers and protective mats, particularly if the seats are still a little damp!

5.4.0 Maintenance puzzles

5.4.1 Cryptograms

A	B	C	D	E	F	G	H	I	J	K	L	M	N	O	P	Q	R	S	T	U	V	W	X	Y	Z
												24					8								

R _ _ _ _ _ _   M _ _ _ _ _ _ _ _ _ _   _ _ _ _ _
8 25 19 14 12 22 21   24 1 12 22 14 21 22 1 22 13 21   4 21 21 6 3

_   _ _ _ _ _ _ _   _ _   _ _ _ _   _ _ R _ _ _ _
1   10 21 18 12 13 7 21   12 22   26 25 25 9   15 25 8 4 12 22 26

_ R _ _ R
25 8 9 21 8

Cryptogram 5.1 Reason for servicing:

5.4.2 Anagrams

Antenna Mice   Canine Meant
Ancient Name   Manic Neaten

Anagrams 5.1 Keep it running: what word did I use to get these anagrams?

Lingo Hips   Poling His
Losing Hip   Nigh Lip So

Anagrams 5.2 Make your car shiny happy car: what word did I use to get these anagrams?

5.4.3 Word search

U	S	B	E	T	N	E	N	O	P	M	O	C	E	L	
T	B	I	P	N	W	N	S	M	N	A	Q	S	L	J	
N	R	V	S	X	G	G	Z	I	L	I	X	S	C	M	
L	A	A	G	S	N	I	H	X	N	N	B	E	I	Q	
U	E	U	N	I	A	H	N	S	U	T	J	W	H	G	
M	Y	H	R	S	P	H	P	E	S	E	M	H	E	G	CHASSIS
G	S	P	G	K	M	E	C	Q	C	N	X	S	V	J	COMPONENT
C	S	H	I	I	C	I	I	G	E	A	R	B	O	X	DESIGN
Q	J	G	L	T	U	U	S	W	M	N	K	B	V	L	ENGINE
F	Q	F	I	F	N	G	C	S	N	C	W	D	R	O	GEARBOX
H	G	O	P	S	H	F	Z	B	I	E	Y	E	K	J	INSPECTIONS
H	N	H	K	Q	C	L	Q	Y	E	O	D	S	T	A	MAINTENANCE
S	X	P	C	L	R	K	K	M	A	E	N	I	X	A	SPRINGS
R	U	O	S	X	M	O	M	F	U	E	G	G	F	X	TRANSMISSION
Q	W	V	F	M	E	O	G	B	S	F	P	N	I	H	VEHICLE

Word search 5.1 Vehicle overview

Remember! There is a multimedia version of this textbook that includes additional images and interactive features: www.automotivett.org

5.4.4 Crosswords

Across

2 A vehicle frame consisting of two side members and a series of cross members (6,6)

5 up Process of making sure a fluid is full to the appropriate level mark (7,2)

7 Structural lower part of a vehicle to which the running gear and body are attached

9 A safety feature of vehicle body that absorbs impacts (7,4)

10 A manual check of suspension dampers (shock absorbers) by pushing down (6,4)

Down

1 Original equipment manufacturer (1,1,1)

3 A vehicle that uses two or more power sources to move the vehicle

4 Imperial speed (1,1,1)

5 Any axis running across a vehicle at right angles to the length

6 A thick sticky lubricant used for bearing and chassis lubrication

8 Rate at which fuel is used (1,1,1)

Crossword 5.1 Vehicle overview and maintenance

CHAPTER 6

Engine Systems

6.1 Principles of engine components and operation

Together with the multimedia resources, this section is ideal material for students working towards the IMI Awards unit EL04, the City and Guilds 3902–011 unit and all other similar foundation or introduction level qualifications.

After successful completion of this section you will be able to show you have achieved these objectives:

- Be able to work safely.
- Know about components of four stroke internal combustion engines.
- Be able to remove and refit simple four stroke engine components.

Automotive Technician Training: Entry Level 3. 978-0-415-72040-3.
© Tom Denton. Published by Taylor & Francis. All rights reserved.

6.1.1 What you must know about engine components and operation

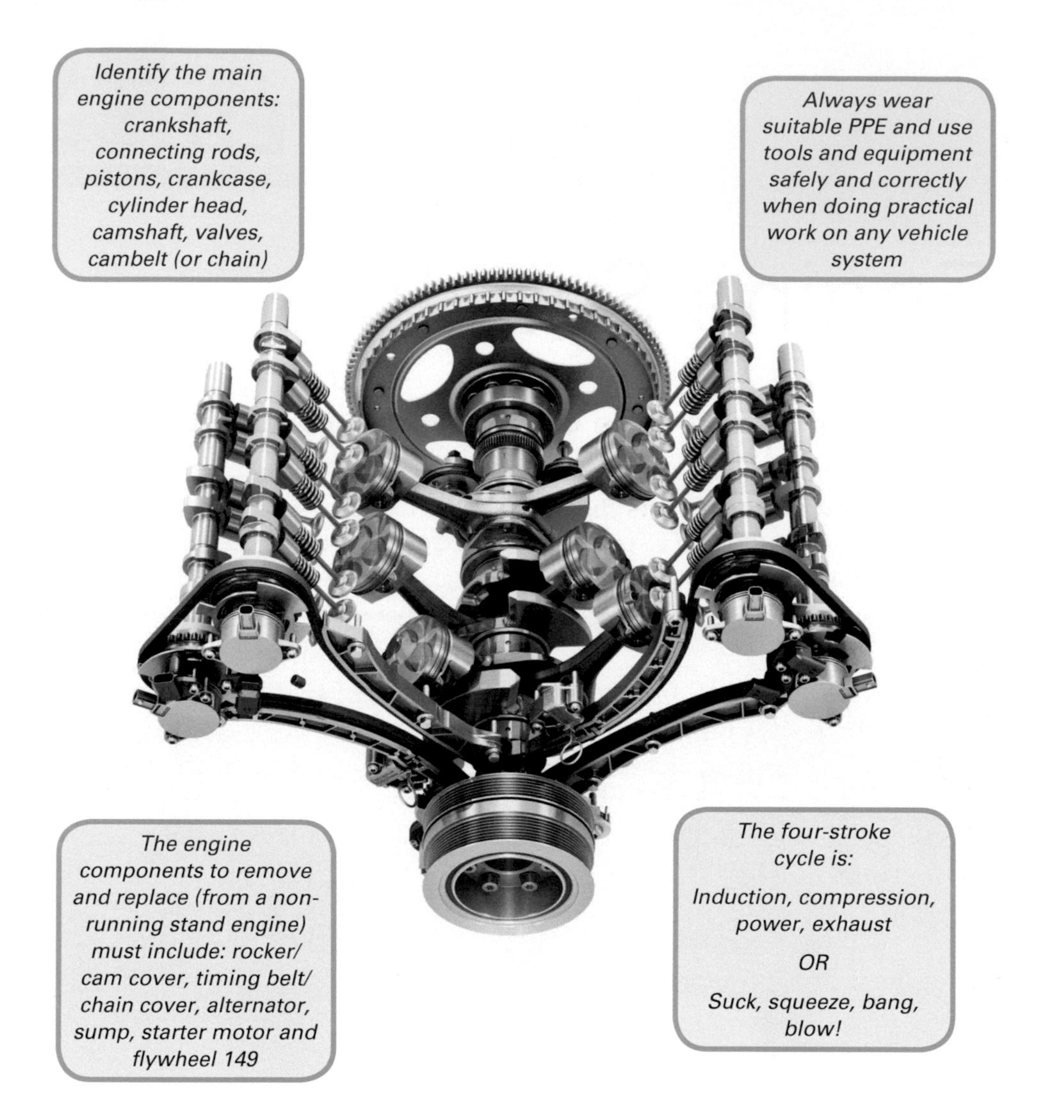

6.1.2 Operating cycles

Introduction The modern motor vehicle engine is a complex machine and the power plant of the vehicle. The engine burns a fuel to obtain power. The fuel is usually petrol or diesel, although liquid petroleum gas (LPG) is sometimes used and specialist fuels have been developed for special purposes such as racing car engines.

Internal combustion engine Motor vehicle engines are known as 'internal combustion' engines because the energy from the combustion of the fuel and the resulting pressure from expansion of the heated air and fuel charge is applied directly to pistons inside closed cylinders in the engine. The term 'reciprocating piston engine' describes the movement of the pistons, which go up and down in the cylinders. The pistons are connected by a rod to a crankshaft to give a rotary output.

Figure 6.1 Internal combustion engine internal components

Air and fuel The fuel is metered into the engine together with an air charge for most petrol engines. Some use injectors that inject directly into the engine cylinder. On diesel engines, the fuel is injected into a compressed air charge in the combustion chamber. In order for the air and fuel to enter the engine and for the burnt or exhaust gases to leave the engine a series of ports are connected to the combustion chambers. The combustion chambers are formed in the space above the pistons when they are at the top of the cylinders. Valves in the combustion chamber at the ends of the ports control the air charge and exhaust gas movements into and out from the combustion chambers.

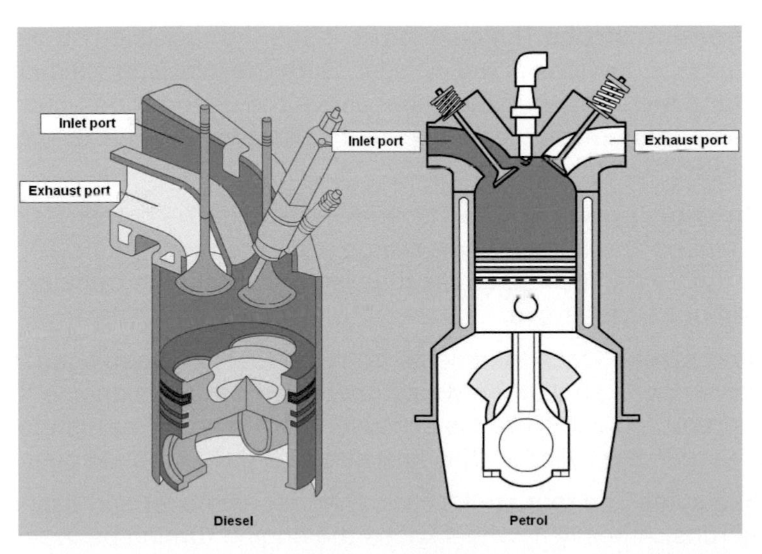

Figure 6.2 Diesel and petrol engine – pistons at BDC before start of intake strokes

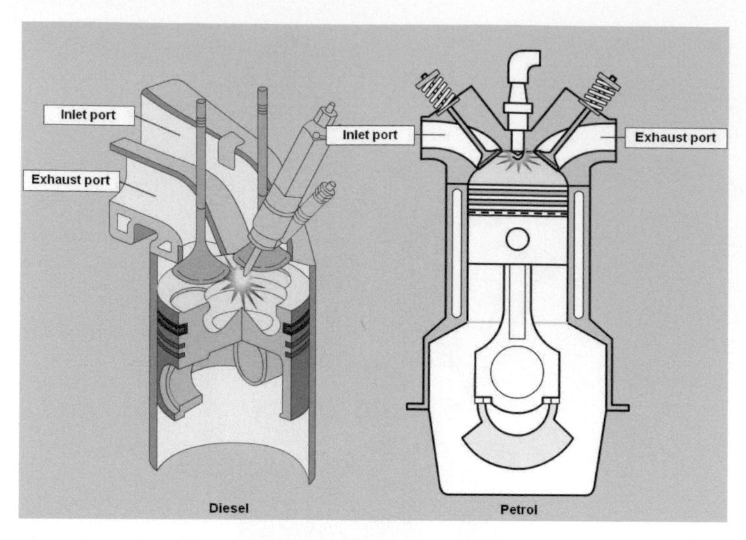

Figure 6.3 Diesel and petrol engine – pistons at TDC before start of the combustion or power strokes

Poppet valves The valves are known as 'poppet' valves having a circular plate at right angles to a central stem that runs through a guide tube. The plate has a chamfered sealing face in contact with a matching sealing face in the port. The valve is opened by a rotating cam and associated linkage. It is closed and held closed by a coil spring.

The four-stroke cycle (or Otto cycle) The opening and closing of the valves and the movement of the pistons in the cylinders follows a cycle of events called the 'Four-stroke cycle' or the 'Otto cycle' after its originator.

The induction/intake stroke The first stroke of the four stroke cycle is the induction or intake stroke. This occurs when the piston is moving down in the cylinder from top dead centre (TDC) to bottom dead centre (BDC) and the inlet valve is open. The movement of the pistons increases the volume of the cylinder so that air and fuel enter the engine.

The compression stroke The next stroke is the Compression stroke when the piston moves upwards in the cylinder. Both the inlet and exhaust valves are closed and the space in the cylinder above the piston is reduced. This causes the air and fuel charge to be compressed, which is necessary for clean and efficient combustion of the fuel.

The combustion/power stroke Towards the end of the compression stroke, the fuel is ignited and burns to give a large pressure rise in the cylinder above the piston. This pressure rise forces the piston down in the cylinder on the Combustion or Power stroke.

The exhaust stroke Once the energy from the fuel has been used, the exhaust valve opens so that the waste gases can leave the engine through the exhaust port. To complete the exhausting of the burnt gases the piston moves upward in the cylinder. This final stroke is called the Exhaust stroke.

Four-stroke cycle The four-stroke cycle then repeats over and over again, as the engine runs. A heavy flywheel keeps the engine turning between power strokes.

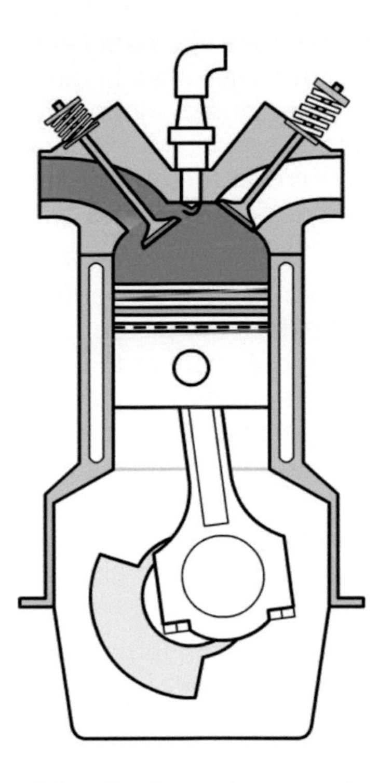

Figure 6.4 Induction – piston moving down

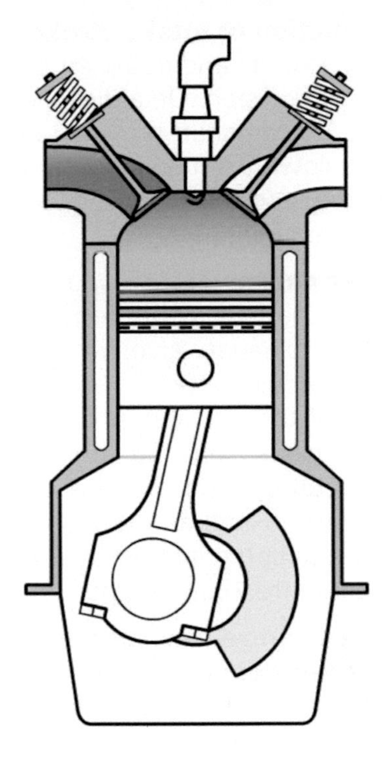

Figure 6.5 Compression – piston moving up

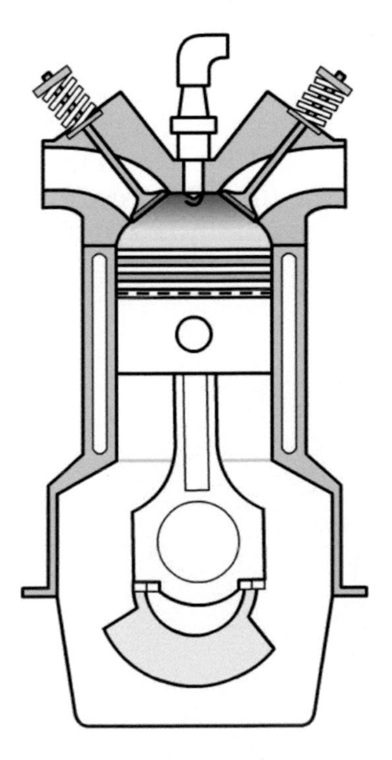

Figure 6.6 Power or combustion – piston is forced down

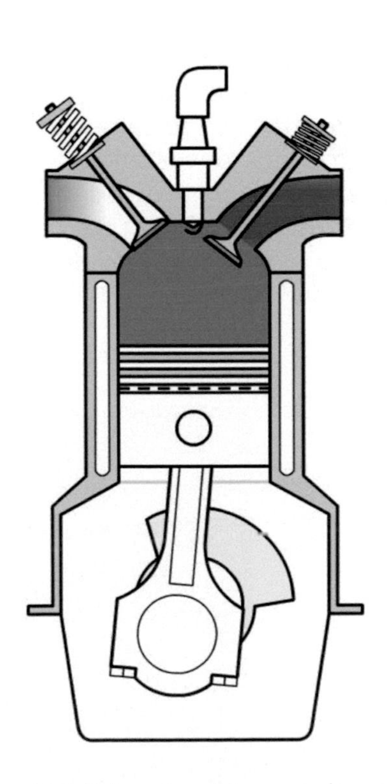

Figure 6.7 Exhaust – piston moving up

The induction or intake stroke On the induction stroke of a petrol engine (most types), air and petrol enters the cylinder so the inlet valve in the inlet port must be open. On a diesel engine, only air enters the cylinder. A rotating cam on the camshaft provides a lifting movement when it runs in contact with a follower. A mechanical linkage is used to transfer the movement to the valve stem and the valve is lifted off its seat so that the inlet port is opened to the combustion chamber.

Cylinder charge The air and fuel charge or air charge can now enter the cylinder. The inlet valve begins to open shortly before the piston reaches TDC. The exhaust valve, which is operated by its own cam in the same way as the inlet valve, is beginning to close as the piston passes TDC at the end of the exhaust stroke. Valve overlap helps clear the remaining exhaust gases from the combustion chamber. The incoming air charge fills the combustion chamber as the last quantity of exhaust gas leaves through the exhaust port. This is known as 'scavenging' and helps cool the combustion chamber by removing hot exhaust gases and gives a completely fresh air charge.

Top dead centre (tdc) and bottom dead centre (bdc) The terms top dead centre and bottom dead centre are abbreviated to 'TDC' and 'BDC' respectively. They are used to describe the position of the piston and crankshaft when the piston is at the end of a stroke and the axis of the piston and crankshaft bearing journals are in a straight line and at 0° (TDC) and 180° (BDC) of crankshaft revolution. To the abbreviations are added the letters 'A' to indicate degrees 'after' TDC or BDC and the letter 'B' to indicate 'before' TDC or BDC.

Crankshaft and camshaft The camshaft rotates once for the two revolutions of the crankshaft during the four stroke cycle. The drive from the crankshaft to the camshaft has a 2:1 ratio produced by the numbers of teeth on the driven and driver gears. Rotational data for the camshaft is usually given as degrees of crankshaft rotation and this should to be considered in relation to the four-stroke cycle. The four-stroke cycle occurring over two full revolutions of the crankshaft has a 720° rotational movement.

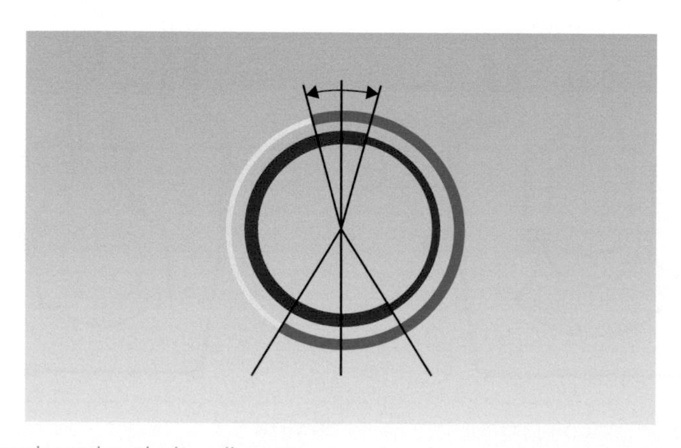

Figure 6.8 Circular valve timing diagram

Use the interactive media search tools to look for pictures and videos to examine the subject in this section in more detail.

Engine locations No matter what design of engine, it has to be positioned in the vehicle. There are various configurations that manufacturers have used in the configuration of their vehicle power trains. The engine can be front, mid or rear mounted and can be installed in-line (along the vehicle axis) or transverse (across the vehicle axis).

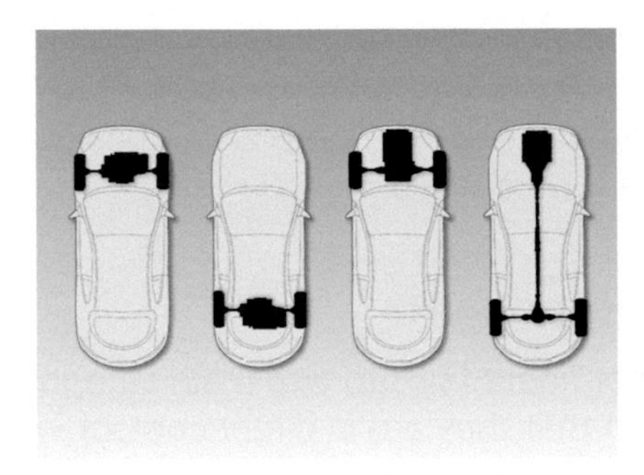

Figure 6.9 Typical positions for the engine

Engine mounting in the vehicle frame The engine mounting system is important as it supports the weight of the engine in the vehicle. In addition, it counteracts the torque reaction under load conditions. The mounting system has to isolate the vehicle from the engine structure borne vibrations. The engine mounts consist of steel plates with a rubber sandwich between to provide the vibration isolation. The mountings have appropriate brackets and fittings to fix to the engine and vehicle frame

Figure 6.10 Engine mountings

Summary This animation shows a view from above of the four-stroke cycle operating in a four cylinder engine. Note the firing order of 1–3–4–2 and how each cylinder runs through the four-strokes: induction, compression, power and exhaust – or, suck, squeeze, bang, blow!

Look back over the previous section and write out a list of the key bullet points.

Answer the following questions either here in your book or electronically.

1 Explain what is meant by 'internal combustion'.
2 Explain the four-stroke cycle.

6.1.3 Cylinder components

Cylinder liners Cylinder liners fall into two main categories, wet and dry. Wet liners are installed such that they are in direct contact with the coolant fluid. They are fitted into the block with seals at the top and bottom and are clamped into position by the cylinder head. Spacers are fitted at the bottom to adjust the protrusion of the liner to achieve the correct clamping force.

Dry liners Dry liners are not in direct contact with the coolant. Generally, they are fitted into the casting mould and retained by shrinkage of the casting via an interference fit. Alternatively, they can be pressed into place in a pre-cast

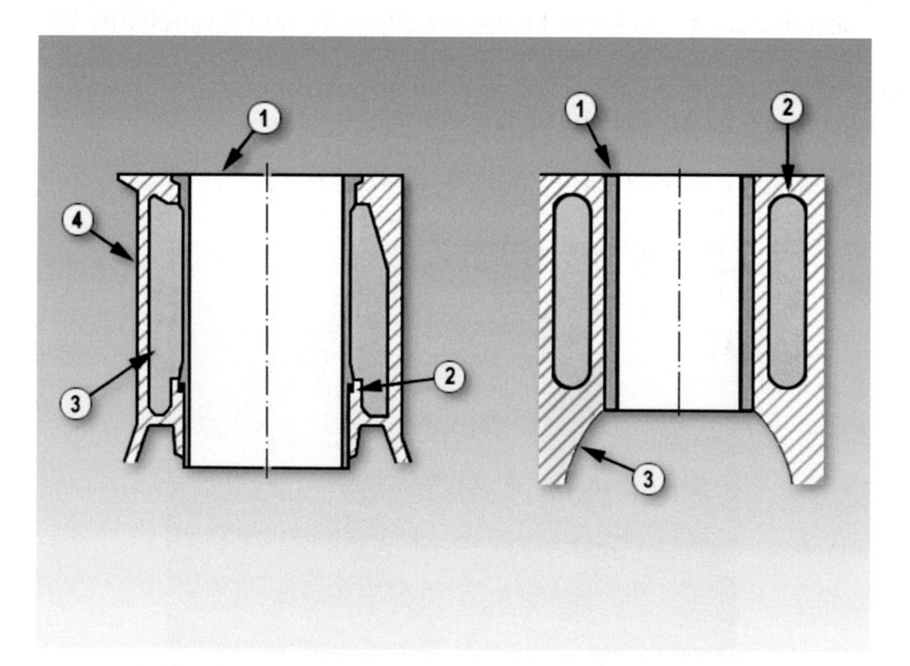

Figure 6.11 Cylinder liners

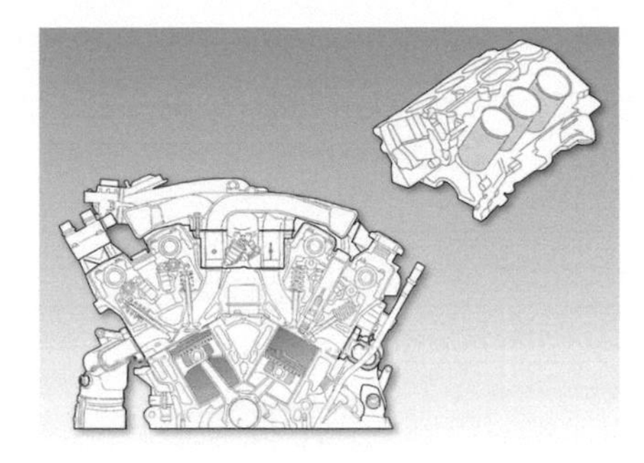

Figure 6.12 Dry liners are not in contact with the coolant

cylinder block. When repairing or reconditioning the engine, The former type can be re-bored whereas the latter type is replaceable.

Replaceable liners Most modern engines have specific treatments applied to the cylinder bores and as such, cannot be re-bored or honed. Replaceable liners mean that the liner and piston assembly can be easily replaced without the need for specialist re-boring equipment. Often, commercial vehicle engines utilise replaceable liners to reduce repair times.

Figure 6.13 Liners

Separate crankcase and block For certain engine applications, separate crankcase and cylinder blocks are utilised. This is not often seen for light vehicles but is quite common on larger engines for commercial vehicles. The cylinder block can be arranged for multiple cylinders or a single block assembly per cylinder.

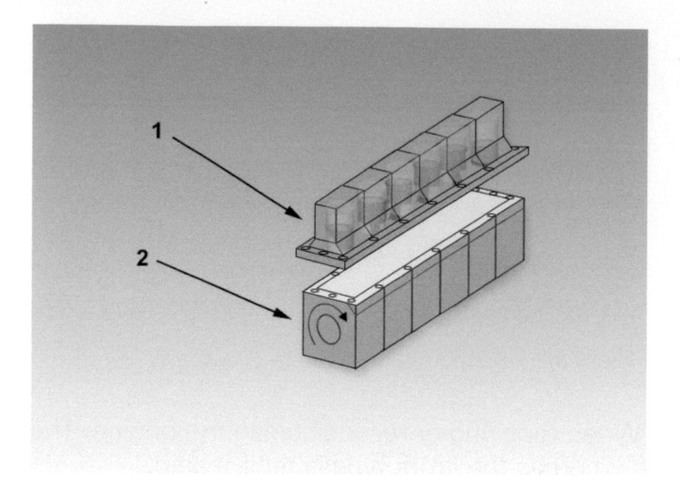

Figure 6.14 Separate components

Answer the following questions either here in your book or electronically.

1 State the difference between wet and dry cylinder liners.
2 State an advantage of each type.

Look back over the previous section and write out a list of the key bullet points.

Construct a word search grid using some key words from this section. About 10 words in a 12 × 12 grid is usually enough.

6.1.4 Valves and valve gear

Overhead valve (OHV) layouts This term is used to describe the evolution of engines from side to overhead valve. It is used to describe the latter, where the valves are located in the cylinder head. The overhead valve engine valve gear is more complex as the valve motion had to be transferred to valves that are facing in the opposite direction when compared to a side valve engines.

OHV valve gear OHV valve gear transfers reciprocating motion from the cam followers and camshaft, to the valves via push rods and a rocker assembly which acts on the valve stems.

Valve clearances Adjustment of the clearance between the rocker and valve stem contact face allows for thermal expansion and correct lubrication. The adjustment is effected via screw adjusters and lock nuts. Hydraulic lifters can be used in place of cam followers to provide self-adjustment of the clearance.

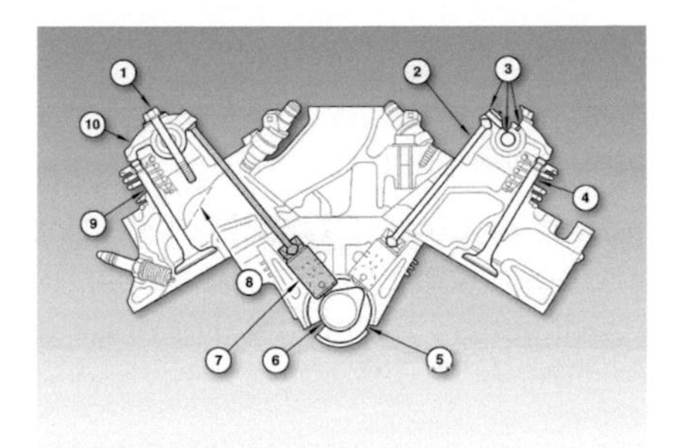

Figure 6.15 Overhead valves (OHV)

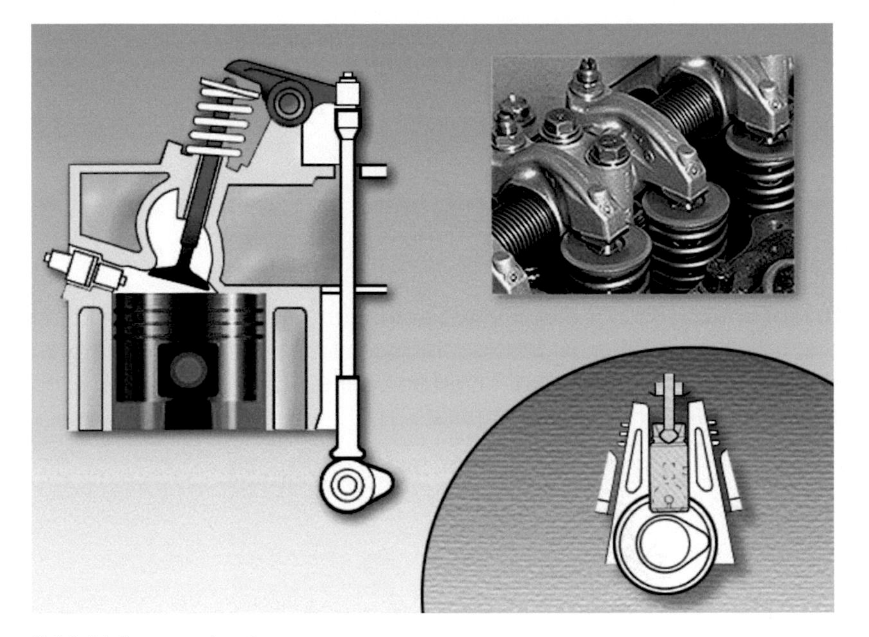

Figure 6.16 Valve mechanisms

Overhead cam (OHC) layout Overhead Cam refers to the position of the camshaft in the engine, that is, it is positioned in the cylinder head. There are a number of designs using direct or indirect mechanisms to convert the

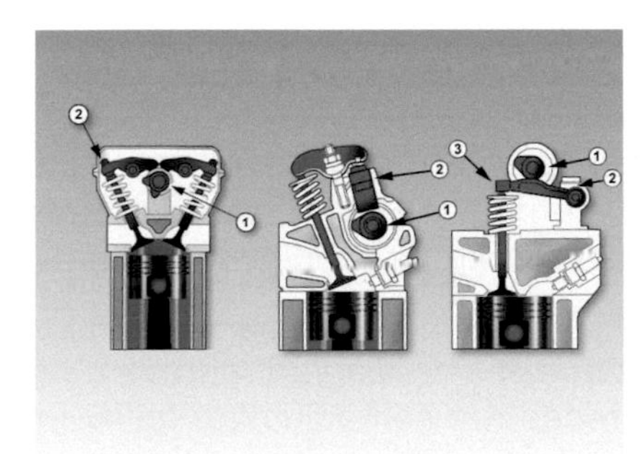

Figure 6.17 Overhead camshaft (OHC) valve operation

Remember! There is a multimedia version of this textbook that includes additional images and interactive features: www.automotivett.org

rotating cam motion into a reciprocating motion, and then transferring this motion to the valve stems. These designs are proposed in order to facilitate a close tolerance in operating clearances

Double overhead camshaft engine (DOHC) A further development is the use of twin or double overhead camshafts. These can use direct or indirect valve actuation and are well suited to multi-valve engine designs, including those with variable valve timing.

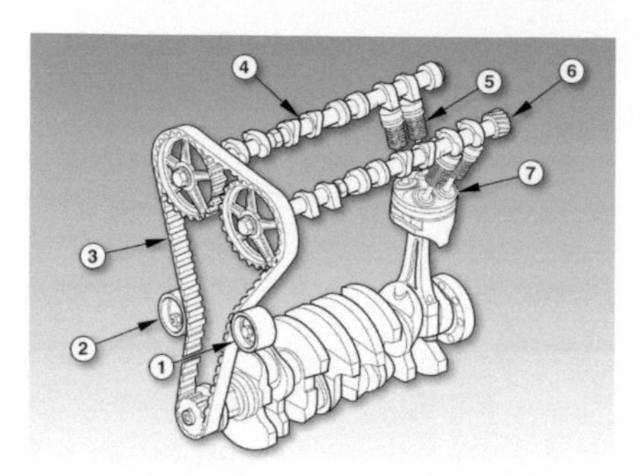

Figure 6.18 Double overhead camshaft engine (DOHC)

Adjustment methods For direct valve actuation with inverted bucket-type followers, adjustment is provided via shims or screw wedge-type adjusters. In addition, hydraulic 'self-adjusting' inverted bucket followers can be employed that require no manual adjustment task.

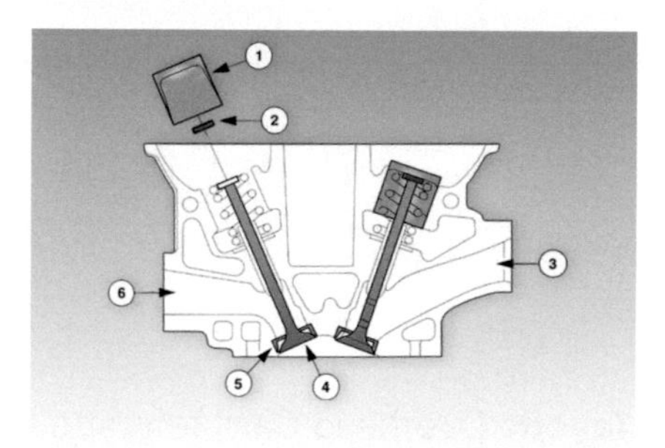

Figure 6.19 Valve components

Rocker arm systems Indirect, rocker arm-type valve actuators incorporate close tolerance adjusters. There are two systems commonly seen, a rocker shaft and pivot stud or, a rocker arm supported on a pedestal at one end and the valve stem at the other, the cam acts between these two points. A hydraulic pedestal can be employed for self-adjustment of the mechanism.

Hydraulic valve adjustment The diagram shows a typical engine oil lubrication circuit that feeds the self-adjusting followers with pressurised oil to maintain the correct valve clearances. Always refer to manufacturers data for the service requirements of the valve train system. Often, special procedures are required when replacing and re-commissioning self-adjusting valve mechanisms, these must be followed to prevent engine damage.

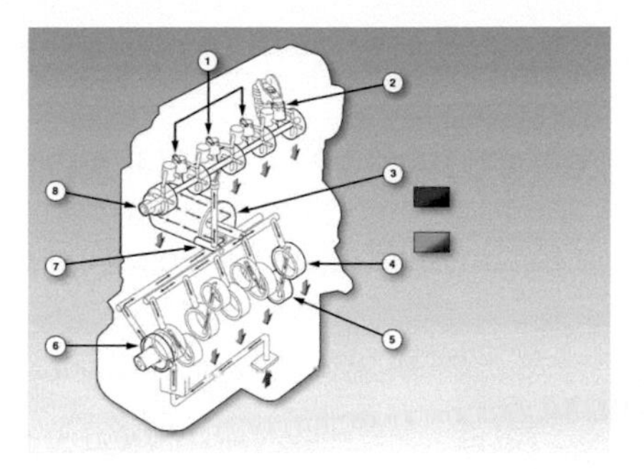

Figure 6.20 Hydraulic components

Look back over the previous section and write out a list of the key bullet points.

Make a simple sketch to show how one of the main components or systems in this section operates.

6.1.5 Engine electrical

Introduction A good supply of electric power is necessary for modern vehicles. The engines require a large current in order to start and many other systems are now electrically powered. Most small vehicles use a 12 V system but it is likely that, in the future, 42 V systems will become standard. This

Figure 6.21 Alternator and starter

is to provide sufficient power for the ever-increasing range of electrical and electronic accessories.

Microprocessors Many components that were once mechanically operated, are now driven by small electric motors and controlled by microprocessors. Total vehicle control, through sensors, electronic control units and actuators, may be common on vehicles sometime in the future.

Battery and charging system function The main function of the battery and charging system is to provide a source of electric power for all the electrical systems on the vehicle. They must be capable of providing the electric power under all operating conditions.

Starting system function The main function of the starting system is to crank the engine at sufficient speed to begin the internal combustion process. This will then allow the engine to run, and to be fully controlled by the vehicle driver.

Figure 6.22 Charging system

Figure 6.23 Starter system

Lead-acid batteries The majority of vehicle batteries are of conventional design, using lead plates in a dilute sulphuric acid electrolyte. This feature leads to the common description of 'lead-acid' batteries. The output from a lead-acid battery is direct current (DC).

Battery chemistry A rechargeable battery is an electrochemical unit that converts an electric current into a modified chemical compound. This chemical reaction can be reversed to release an electric current. The modified chemical compound in the battery stores energy, which is available as electricity when connected to a circuit.

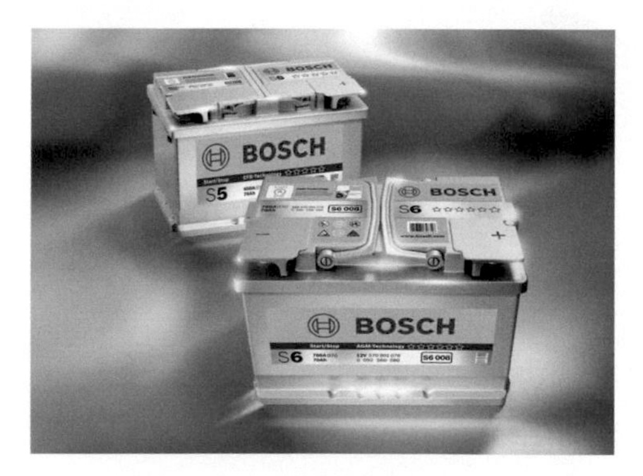

Figure 6.24 Vehicle batteries

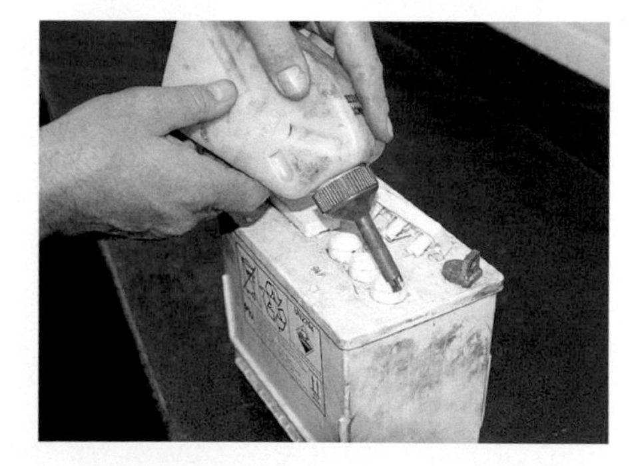

Figure 6.25 Open cell battery – with removable caps

Topping up the battery Only water is lost from the battery and therefore only water should be used for topping up. Any contaminants will affect the chemical reactions in the battery and, therefore, the performance. Only distilled or specially produced topping up water should be used. Tap water is not suitable for topping up a battery. Acid should never be used, as this will strengthen the acid solution and alter the chemical reactions.

Note: Not many batteries can be topped up now but there are a few out there!

Look back over the previous section and write out a list of the key bullet points.

Construct a crossword puzzle using important words from this section. Hint: use the ATT glossary where you can copy the words and definitions (clues!). About 20 words is a good puzzle.

6.2 Routine cooling and lubrication system checks

Together with the multimedia resources, this section is ideal material for students working towards the IMI Awards unit EL08, the City and Guilds 3902–001/002 units and all other similar foundation or introduction level qualifications.

After successful completion of this section you will be able to show you have achieved these objectives:

- Be able to work safely.
- Know about cooling systems.
- Be able to check a cooking system.
- Know about engine lubrication systems.
- Be able to check a lubrication system.
- Know about disposal of harmful substances.

6.2.1 What you must know about lubrication and cooling

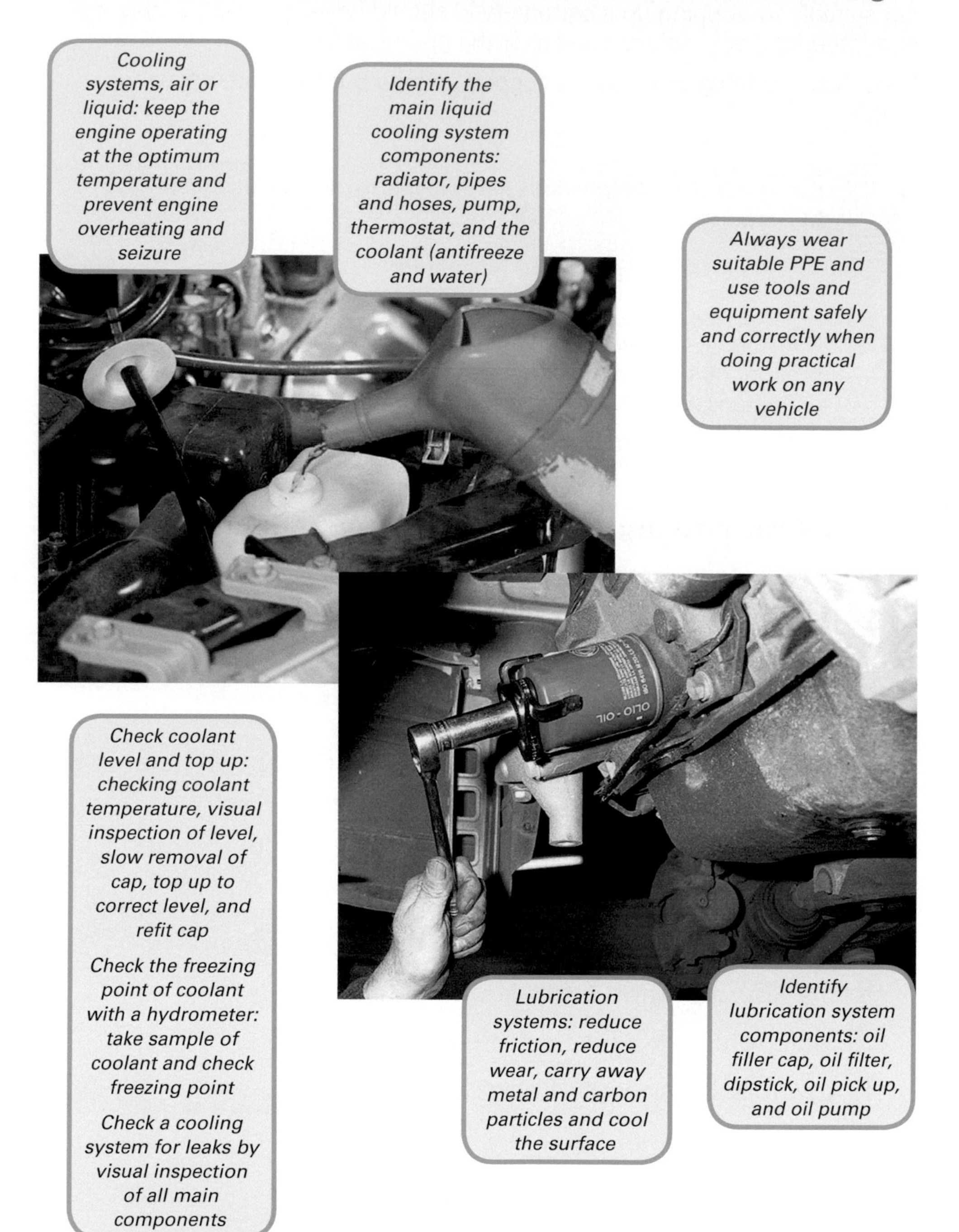

6.2.2 Cooling introduction

The main function of the cooling system is to remove the heat from the engine, particularly around the cylinder walls and the combustion chambers. This should occur under all operating conditions, including the extremes of very hot weather, hard driving and high altitude.

Emissions The engine cooling system on a modern motor vehicle has to play its part in keeping exhaust emissions to a minimum. During cold start and warm-up, an engine requires a rich petrol-to-air mixture to run smoothly. Because a cold engine produces high levels of unwanted exhaust emissions, a rapid warm-up is needed to keep emissions to a minimum. The 'normal' running engine-coolant temperature is maintained at about 90°C, which gives an engine temperature enabling clean combustion.

Emission control The control of emissions is achieved by controlling the upper-cylinder and combustion-chamber temperatures, resulting in the efficient and clean combustion of the fuel. A further reduction in harmful exhaust emissions is achieved by keeping the warm-up time to a minimum.

Figure 6.26 Engine cross-section

Warm-up time Warming up to the optimum temperature as quickly as possible is important, not only because it helps to reduce exhaust emissions, but also helps to prevent the formation of water particles in the combustion chamber and exhaust when the engine is cold. Any water that does not evaporate can enter the engine and contaminate the engine oil, or remain in the exhaust system and cause premature corrosive damage.

Water jacket The water jacket is cast into the cylinder block and cylinder head. Casting sand is used to shape the inside, or core, of the casting for the water passages. The sand is removed after casting through a series of holes in the sides, ends and mating faces of the cylinder block and head.

Water passages The holes in the sides and ends of the block and head are machined to provide accurate location for core plugs that complete the outside water tightness of the water jacket. The holes in the mating faces are aligned to allow coolant flow from the cylinder block to the cylinder head. These components are also machined for the fitting of the water pump and a water outlet to the radiator.

Coolant flow The internal designs of the head and block vary to give different coolant-flow patterns. An even flow to all areas of the engine is very important. The main areas where cooling is needed are around the combustion chambers and the upper-cylinder walls.

Figure 6.27 Engine block with core plugs

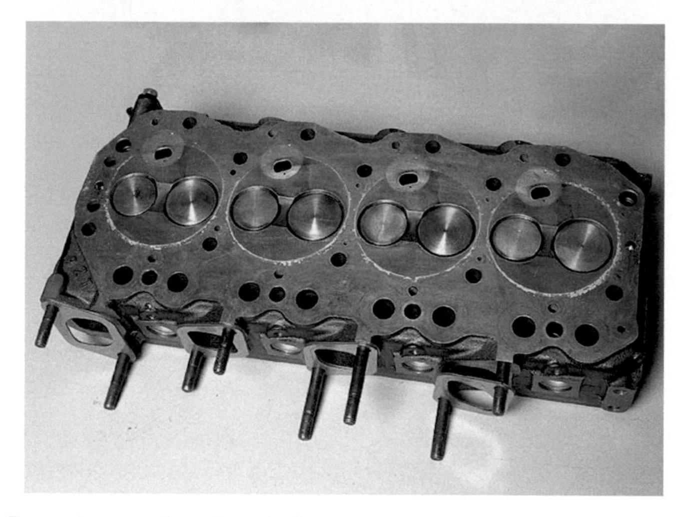

Figure 6.28 Core plugs and coolant holes

Cylinder head The need for inlet ports, exhaust ports and valves makes cooling of these regions difficult. These areas are prone to cracking and other deterioration from overheating, freezing, and the use of incorrect or old, antifreeze solutions.

Air cooling Some older engine designs used an air-cooling system. Modern engines use water-cooling because this is capable of giving the precise engine temperature control needed for exhaust-emission regulations.

Bypass system Recent developments in coolant circulation give improved control of engine temperature. Mixing cold and hot water as it enters the engine achieves this, as opposed to the cold fill of earlier systems. Both the old and new systems are covered in this learning programme.

Figure 6.29 Air-cooled system

Figure 6.30 Water-cooled system

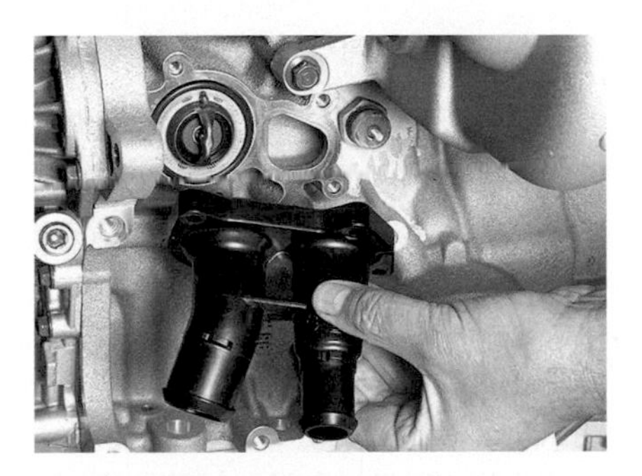

Figure 6.31 Ford engine coolant ports

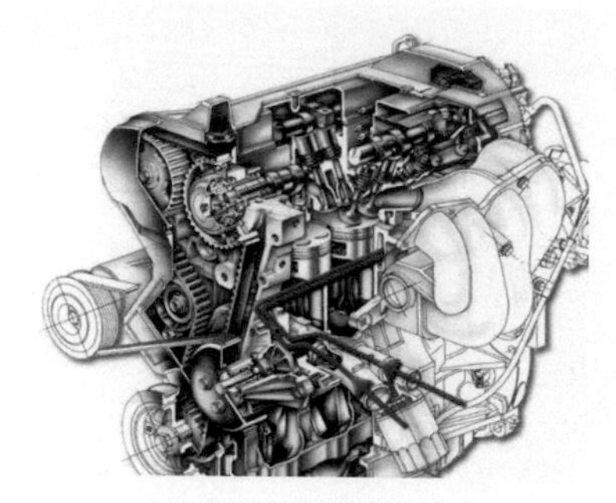

Figure 6.32 Ford engine coolant circuit

Service life Cooling-system components must have a service life that is comparable with the engine mechanical components. However, some are subject to wear and natural deterioration and need to be replaced at scheduled-service intervals.

Coolant The coolant must be able to resist freezing and boiling. Contamination and corrosion of engine and cooling-system components must be kept to a minimum.

Expansion Components expand with heat and, at high temperatures, this expansion can cause seizure, burning of pistons and valve seats. High temperature would also produce rapid deterioration of the engine oil.

Overheating A result of overheating is a change in the nature of the combustion process. The combustion time reduces which, in turn, leads to a rapid rise in the pressure and force acting on the piston crown, connecting rod and crankshaft. A 'pinking' sound may be heard and premature failure of these components is likely. There is also an increase in temperature to a point at which high levels of nitrogen oxides are produced and these are harmful to the environment.

Figure 6.33 Thermostat

Cooling-system design Cooling systems are designed to maintain engines at an optimum temperature. This allows the design of components that expand on heating to form very tight fits and running tolerances. The adjustment of ignition and fuel settings are equated to the optimum temperature required for the clean and efficient combustion of fuel.

Air cooling Air-cooled systems have the air stream passing directly over the cylinder heads and cylinders to remove heat from the source. Fins are cast into the cylinder heads and cylinders to increase the surface area of the components, thus ensuring that sufficient heat is lost.

Liquid cooling Liquid-cooling systems use a coolant to carry heat out of the engine and dissipate the heat into the passing air stream. The liquid coolant is contained in a closed system and is made to circulate almost continuously by the impeller on the water pump. Heat is collected in the engine and dissipated from the radiator into the passing air stream.

Look back over the previous section and write out a list of the key bullet points.

Create a mind map to illustrate the important features of a component or system in this section.

6.2.3 Cooling components

Coolant The coolant is a mixture of water and antifreeze. The antifreeze is needed to prevent water expanding as it freezes. The force from that expansion would be sufficient to cause engine cylinder blocks and radiators to burst apart.

Antifreeze Sufficient antifreeze is needed for the climate in which the vehicle is operated. Modern antifreeze formulae are also designed to give year-round protection by increasing the boiling point of the coolant for hot-weather use.

Figure 6.34 Cooling fan and radiator

Thermostat Liquid-cooling systems traditionally used a thermostat in the outlet to the top hose to control engine temperature.

A thermostat is a temperature-sensing valve that opens when the coolant is hot and closes as the coolant cools down. This allows hot coolant to flow from the engine to the radiator where it cools down and returns to the engine. The cooled coolant in the engine acts on the thermostat and it closes.

Coolant flow The coolant re-heats in the engine and the thermostat opens and the cycle of hot coolant flow to the radiator and cool coolant returning to the engine repeats itself. Although this system provides a reasonably effective method of engine-temperature control, it does produce a fluctuating temperature. However, a steady temperature is required for very clean and efficient combustion.

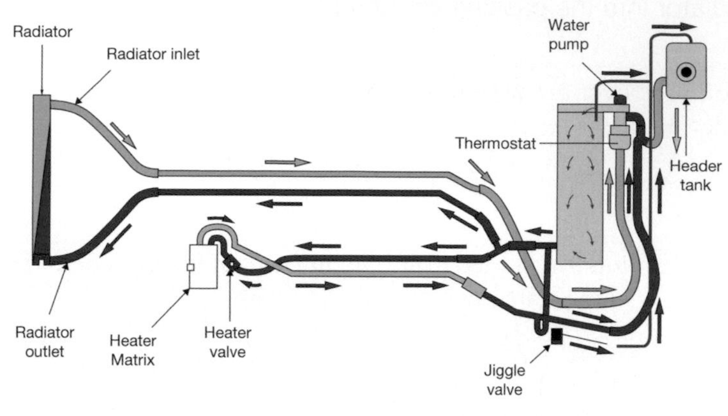

Figure 6.35 Cooling system

Bypass-mixing cooling system Modern engine design is moving towards a system with the thermostat in the radiator bypass channel. When the thermostat opens, it allows cold water from the radiator to mix with the hot-water flow in the bypass as it enters the water pump. This system provides a steady engine temperature and prevents the fluctuating-temperature cycle of the earlier system. The modern system is shown here with arrows indicating the coolant flow.

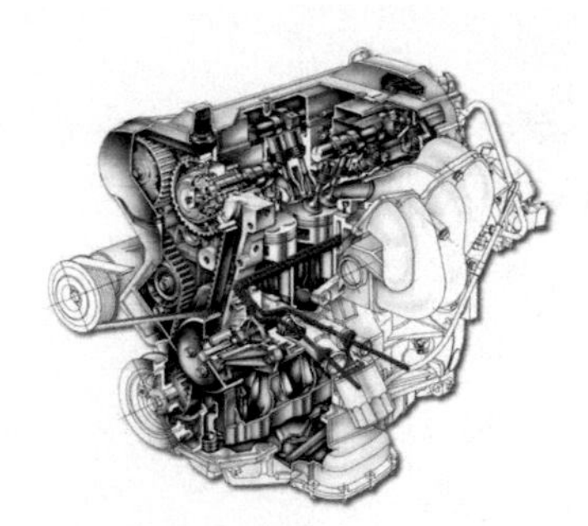

Figure 6.36 Coolant flow

Heat distribution The heat distribution within the engine needs to be controlled. The temperature around all cylinders and combustion chambers should be identical. The heat removed by the cooling system has, therefore, to be consistent for all areas of the engine. All modern engines have a fairly rapid coolant circulation within the engine so that an even temperature distribution is achieved.

Water (coolant) pump The water (or coolant) pump draws the coolant through a radiator bypass channel when the engine is cold and from the radiator when the engine is hot. The impeller on the water pump drives the coolant into the engine coolant passages or water jacket. Water-jacket passages are carefully designed to direct the coolant around the cylinders and upwards over and around the combustion chambers.

Figure 6.37 Water-pump action

Coolant density The density of coolant falls as it heats up and, as the temperature approaches boiling point, bubbles begin to form. These bubbles can create areas in the water jacket where the coolant is at a lower density and the actual mass of coolant in those areas is reduced. The reduced mass of coolant therefore cannot effectively absorb heat efficiently in order to cool the engine.

Another problem of poor heat transfer and lowered coolant density occurs when the rapid flow of coolant into and out of restrictions in the water jacket induces a phenomenon known as 'cavitation'. This results in localised drops in pressure and density in the coolant.

Heat distribution The two causes of localised coolant-density change – bubble formation and cavitation – can seriously affect the performance of the cooling system. This is because an even heat distribution around the cylinders and combustion chambers is not maintained. To overcome these problems, all liquid cooling systems are pressurised. When hot, most modern systems have an operating pressure equivalent to about one atmosphere (1 bar, or 100 kPa).

Expansion The pressure is obtained by restricting the loss of air above the coolant in a radiator header tank or an expansion tank. As coolant heats up it expands. If the air above the coolant has less space to occupy, and it cannot immediately escape, it increases in pressure.

Radiator Pressure Cap A pressure-sensing valve in the radiator cap allows this higher pressure to escape but retains the operating pressure.

Remember! There is a multimedia version of this textbook that includes additional images and interactive features: www.automotivett.org

Figure 6.38 Expansion tank

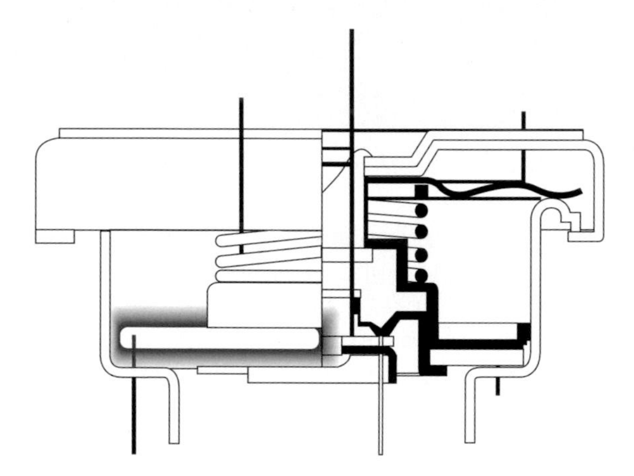

Figure 6.39 Radiator pressure-cap details

Pressure-cap vacuum valve As the engine cools down, the coolant contracts and the pressure drops. A vacuum valve in the pressure cap allows air to return to the system. This prevents depressurisation below atmospheric pressure and also the risk of the inward collapse of components. An early sign of the failure of this valve to open is a top hose that has collapsed.

The pressure in the system acts on the coolant to increase the density, which would otherwise have fallen without the increase in pressure. This helps to reduce the risk of cavitation and to increase the boiling point of the coolant under pressure. The advantages are a more efficient cooling system with a higher safe operating temperature. It can also be used at high altitudes without the need for modification.

Summary A cooling system is needed to prevent engine damage caused by overheating. It also helps to reduce emissions by shortening the engine warm-up time. Heat is used from the cooling system to operate the heater.

Look back over the previous section and write out a list of the key bullet points.

Use the interactive media search tools to look for pictures and videos to examine the subject in this section in more detail.

6.2.4 Friction and lubrication

All types of vehicle engines incorporate metal parts that have to rub against each other, thus causing friction which creates stress, wear and heat, e.g., cylinders in cylinder liners. These parts require lubrication to prevent the wear, keep the surfaces clean, and help to remove the heat.

Lubrication is achieved by separating the metal surfaces with a film of oil or grease. Thus the lubricating oil in the engine has traditionally been seen to have these three functions: separation, cleaning and cooling.

Oil circulates throughout the parts of an engine under pressure produced by a mechanical pump.

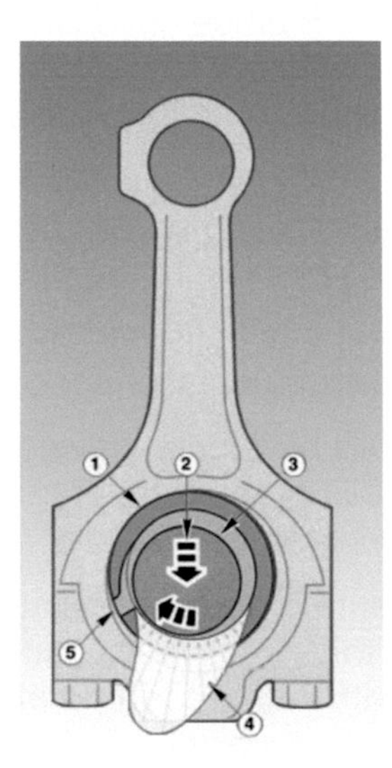

Figure 6.40 Con rod big end lubrication

Another important property of a lubricant is oiliness, which can be described as the ability to adhere to the surface of materials and maintain separation of the rubbing surfaces without breaking down. This type of lubrication is called boundary lubrication and occurs in all engines during starting and before the pumped oil feed is established.

Environmental regulations Modern engines must conform to environmental regulations and modern engine oils are an important component in helping to achieve this. Engine oil producers are responding with new blends, additives and synthetic oils to enable extended service intervals, improved wear protection, greater engine cleanliness, sludge inhibition, higher speeds and temperatures and lower oil consumption. These oils also contribute to improved performance, economy and environmental concerns about hydrocarbon emissions into the atmosphere. They are compatible with oil-seal materials so that leakage is reduced. They also have strict limits on volatility so that vapours do not escape into the atmosphere.

Look back over the previous section and write out a list of the key bullet points.

Use a library or the interactive web search tools to examine the subject in this section in more detail.

6.2.5 Oils and specifications

Most engine oils are refined from crude oil to which are added viscosity index enhancers, reduced-friction enhancers, anti-oxygenates, sludge, lacquer and corrosion inhibitors and cleaning agents for carbon, acids and water.

Early specifications for engine oils defined just the physical data. New oils, which have to meet environmental and engine-performance requirements, are given specification code letters to indicate the performance level.

Synthetic and semi-synthetic oils Synthetic and semi-synthetic oils have improved performance for environmental or special purposes.

Multigrade oils Multigrade oils have been developed in order to modify the viscosity index and give thin oils at low temperatures that do not become excessively thin at higher temperatures.

Viscosity Viscosity is a measure of an oil's resistance to flow, i.e., if thin, the oil will flow more easily than thicker oil. A viscosity index is the measure of a change in an oil's flow rate with a rise in temperature. The higher the viscosity index, the smaller the change in viscosity.

Manufacturer's recommended viscosity ratings generally reflect the lowestv temperature at which the vehicle is being used and may be different for summer and winter use. The viscosity rating is not an indicator of oil quality but of oil flow under particular conditions. There are some low-grade oils that carry recommendations that limit the use of the vehicle, particularly for high engine speeds, loads and long journeys. Good-quality oils will be labelled with at least the API and ACEA service ratings.

Modern engine oil specifications are based on SAE (Society of Automotive Engineers) viscosity ratings, API (American Petroleum Institute) service ratings and other properties defined by classifications laid down by organisations such as ACEA (Constructeurs Européens d'Automobiles) and the earlier CCMC (Comité des Constructeurs d'Automobile du Marché Commun) for European vehicles.

Figure 6.41 Common specifications (SAE is now almost universal)

Recommended oil grades Most engine and vehicle manufacturers list the SAE, API and other classifications for engine oil for their vehicles. They frequently list oil-producer preferences, which give an indication of the co-operation that has been given by the oil producer in the design and development of the engine. Some manufacturers produce their own oils formulated specifically for their vehicles.

Figure 6.42 Two stroke oil

Engine oils are not normally biodegradable and should not be allowed to enter the environment either as vapour or liquid. Total loss lubrication systems used on small two-stroke engines, such as those on motorbikes and outboard motors, use a 'petroil' mixture of petrol and a specially formulated biodegradable oil. Other types of oil should not be used.

Look back over the previous section and write out a list of the key bullet points.

Use a library or the interactive web search tools to examine the subject in this section in more detail.

6.3 Introduction to spark ignition fuel systems

Together with the multimedia resources, this section is ideal material for students working towards the IMI Awards unit L119, the City and Guilds 3902–004 unit and all other similar foundation or introduction level qualifications.

After successful completion of this section you will be able to show you have achieved these objectives:

- Be able to work safely.
- Know the components of spark ignition fuel systems.
- Be able to change air filters.
- Be aware of environmental considerations.

6.3.1 What you must know about spark ignition fuel systems

Always wear suitable PPE and use tools and equipment safely and correctly when doing practical work on any vehicle

Appropriate ways to dispose of waste products in accordance with environmental guidance: disposal of used air filters, disposal of contaminated or spilt fuel, clearing up spillages and disposal of absorbent materials

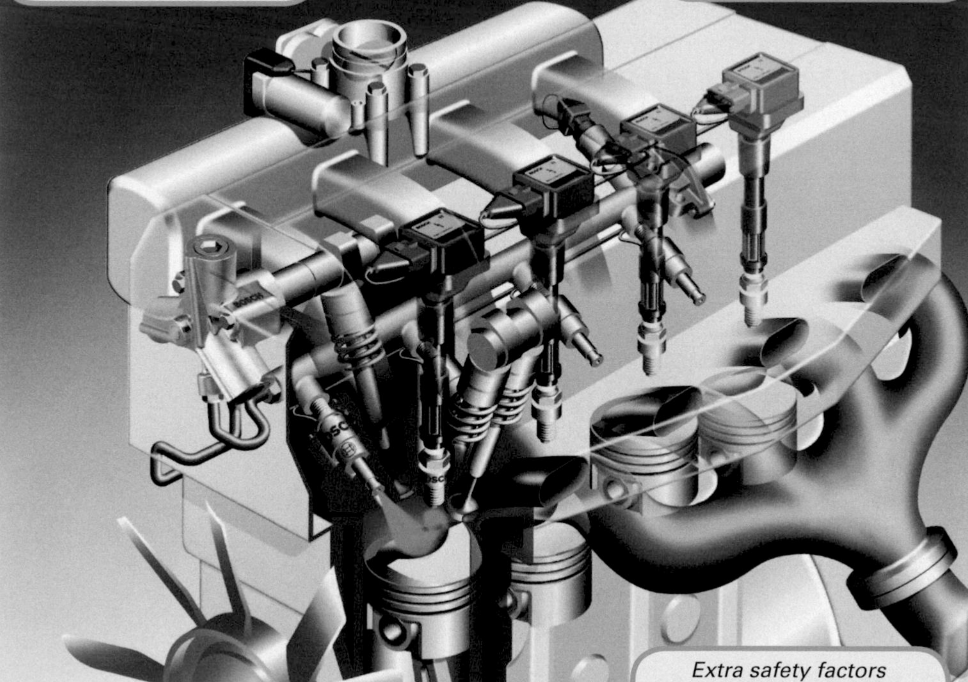

The major parts of the fuel system of spark ignition engines: fuel tank, lines, filter, pressurising system (pump), metering system, delivery system, air intake and filtration

Extra safety factors to be considered when working with fuel systems: fire precautions, exhaust fumes when running an engine in a workshop, handling and disposing of materials, preventing ingress of dirt, moisture and foreign matter

6.3.2 Fuel supply

Fuel supply The fuel on the vehicle is held in a tank fitted in a safe position. Recent construction legislation requires that the tank is unlikely to be ruptured in a vehicle collision. The positioning and protection of the tank is considered at the design stage of the vehicle and tested during development. The tank is fitted with a filler neck and pipe work from the filler cap to the tank. Also fitted are the outlets to the atmospheric vent or evaporative canister and the fuel fed and return pipes to the engine. The fuel gauge is located in the fuel tank. Fuel supply and return lines are made from steel pipes, plastic pipes and flexible rubber joining hoses depending on application and the type of fuel used.

Fuel pumps A pump to supply fuel to the engine is fitted into or near to the tank on petrol injection vehicles. On carburettor vehicles, a mechanical lift pump is fitted to the engine and is operated by a cam on the camshaft or crankshaft, or an electric pump is fitted in the engine compartment. Diesel-engined vehicles using a rotary fuel injection pump, may use the injection pump to lift fuel from the tank. Alternatively, they may have a separate lift pump similar to the ones used on carburettor engines. A separate priming pump fitted in the fuel line may also be used.

Figure 6.43 Fuel injection pump

Mechanical lift pump This diagram shows the operation of a typical mechanical fuel lift pump. Refer to vehicle manuals for details of specific fuel pumps. Modern pumps are sealed units and have to be replaced as a complete unit if they become defective.

Figure 6.44 Mechanical fuel pump

Look back over the previous section and write out a list of the key bullet points.

Create a mind map to illustrate the important features of a component or system in this section.

6.3.3 Petrol injection

Electronic fuel injection (EFI) systems Electronic petrol injection systems have been in use for many years, first on expensive and sports vehicles and now standard equipment on most vehicles. The tougher standards of exhaust emission regulations have made the use of microelectronic control systems for fuel delivery a virtual necessity. There are many different manufacturers of electronic fuel systems and this programme covers the main points of the systems.

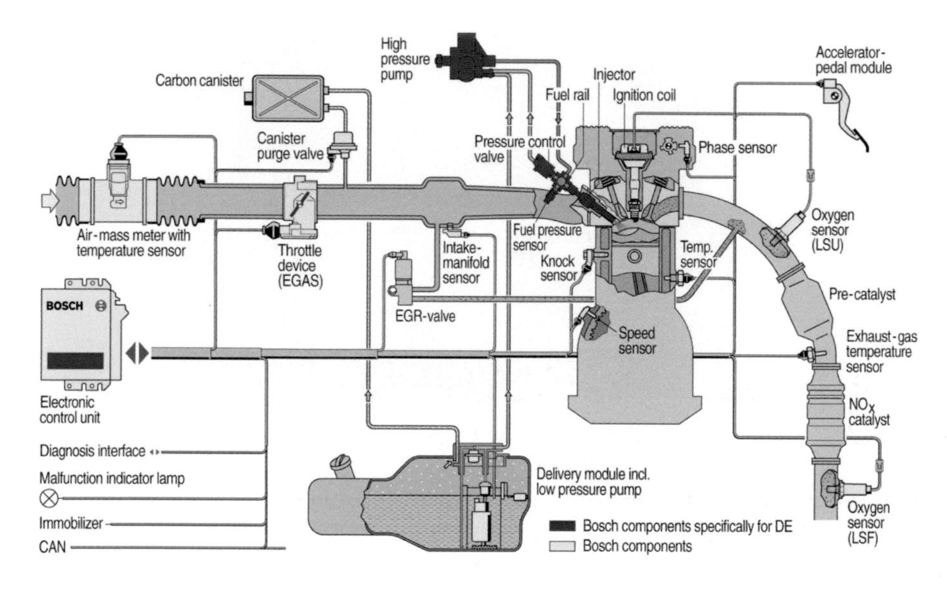

Figure 6.45 EFI system

Electronic control unit At the heart of electronic fuel injection (EFI) systems is the fuel control or electronic control unit (ECU) with a stored map of operating conditions. Electronic sensors provide data to the microprocessor in the ECU, which calculates, and sends the output signals to the system actuators, which are the fuel pump, fuel injectors and idle air control units. The ECU will also switch some of the exhaust emission and auxiliary system components.

Fuel injection methods Electronic fuel injection (EFI) systems are named by the position and operation of the fuel injectors. There is a range of throttle body injection (TBI) systems. They are also known as single point (SPI) or central point (CPI) systems. However they are named, the injector is positioned in a housing fitted on the inlet manifold. This is the same position as the carburettor was traditionally fitted.

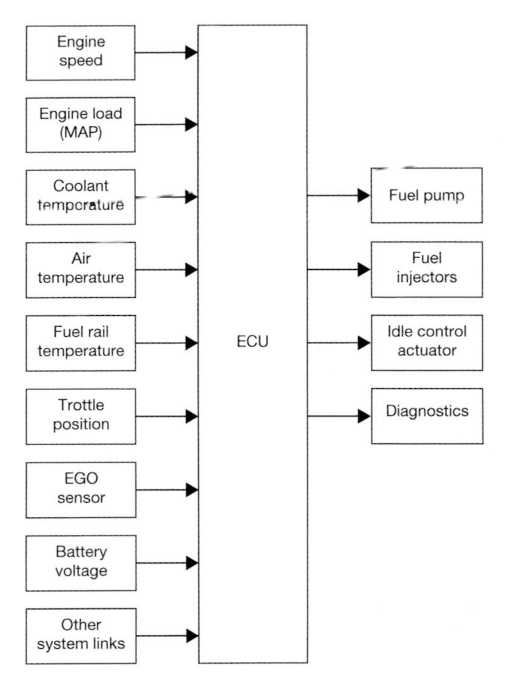

Figure 6.46 ECU with example inputs and outputs

Gasoline direct injection (GDi) A recent development has been the Introduction of a direct injection petrol engine where the fuel is injected into the combustion chamber.

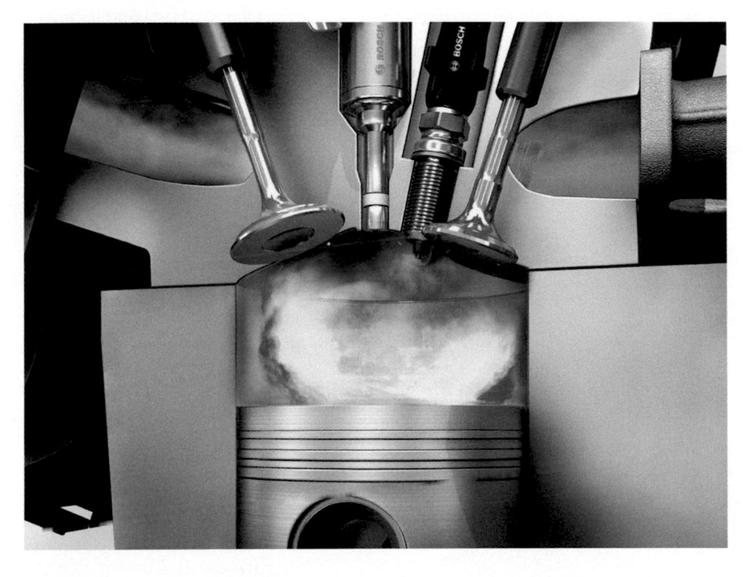

Figure 6.47 GDi (Source: Bosch Media)

Engine control module Modern petrol injection systems are linked to the ignition systems and are controlled by an engine control module (ECM). The latest developments have all electronic systems linked to form a power train control module (PCM). This is also described as a vehicle control module

(VCM). All modern fuel injection systems have closed loop electronic control using an exhaust gas oxygen sensor. For clarity, each electronic control unit will be referred to as an ECU.

Inputs and outputs The components for any electronic fuel injection system can be divided into four groups: The air supply components, the fuel supply components, the electronic control unit (ECU), together with the power supply and system harness and the sensors which provide data to the ECU.

Figure 6.48 EFi components

Air supply components The air supply components consist of ducting and silencing components between the air intake and the inlet manifolds. This will also include an air filter, a throttle body, throttle plate assembly and idle control components. The air supply components must provide sufficient clean air for all operating conditions. The air flow into the engine would be noisy and unbalanced between cylinders without the use of resonators and plenum chambers. A plenum chamber is a large volume air chamber that can be fitted either in front of or behind the throttle plate housing.

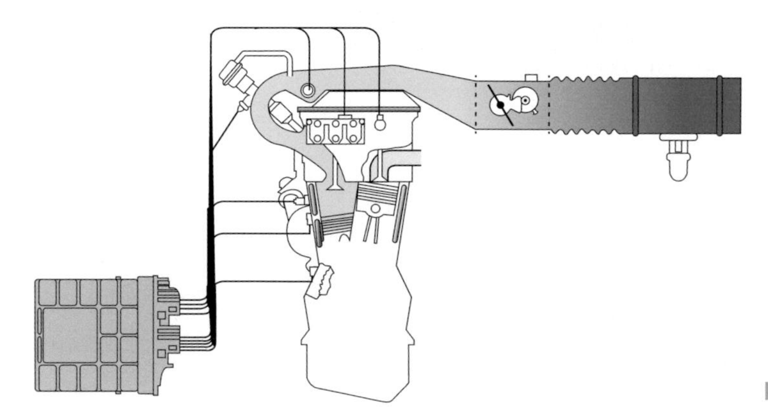

Figure 6.49 Air supply

Air filter Air filters on most modern petrol-engined vehicles consist of a plastic casing with a paper filter element. Air flow into the filter is upwards so that dust and dirt particles drop into the dust chamber, or is rotary so that dust and dirt is thrown out before the air enters the engine. Crankcase ventilation

Figure 6.50 Air filter on a modern vehicle

and the air supply or pulse air exhaust emission systems, are also connected to the filter assembly.

Throttle body The throttle is a conventional circular plate in an air tube. For fast idle and warm up, an auxiliary air valve is fitted to bypass the throttle plate, or an electromechanical link is made to the throttle plate spindle. An auxiliary air valve, idle air control valve (IAC) or idle speed control valve (ISC) is operated from signals from the ECU.

Figure 6.51 Throttle body assembly

Idle speed Control Idle speed control can also be provided by direct action onto the throttle spindle. Electric solenoids or stepper motors are used for this method of control. The solenoids can be single position or multi-position types and can be used for not only cold start and warm up control but also to open the throttle when high load systems, such as the air conditioner, are switched on. Stepper motors give graduated positions depending on the supply current to a number of electric windings. Sensors in the idle control mechanisms provide feedback signals to the ECU to provide data on operation and position.

Fuel supply The fuel supply from the fuel tank to the injector valves for all electronic systems follows the same basic layout. The delivery of fuel at the

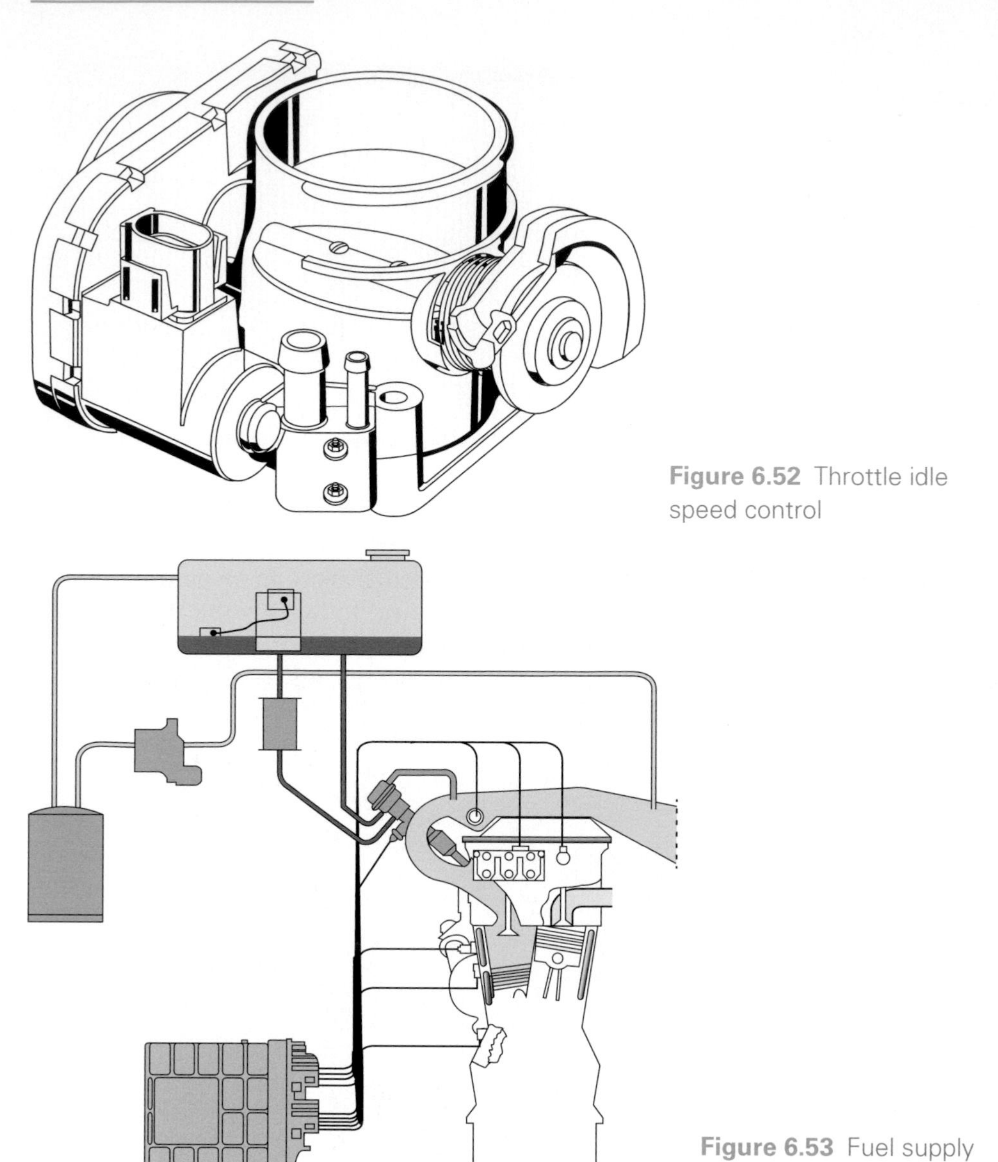

Figure 6.52 Throttle idle speed control

Figure 6.53 Fuel supply components

injector valves is also based on a similar function for all systems. A basic layout of fuel supply components is shown here. A fuel pump is fitted either in, or close to, the fuel tank. A fuel filter is fitted in the delivery fuel lines from the tank to the fuel rail. A fuel pressure regulator is fitted on either the housing for throttle body injector systems, or the fuel rail for port fuel injection systems. The return fuel lines run from the pressure regulator to the fuel tank.

Fuel pump The fuel pump is a roller cell pump driven by a permanent magnet electric motor. Fuel flows through the pump and motor but there is no risk of fire as there is never an ignitable mixture in the motor. The delivery pressure is set by a pressure relief valve, which allows fuel to return to the inlet side of the pump, when the operating pressure is reached. There is a non-return valve in the pump outlet. Typical delivery pressures are between 300 and 400 kPa (3 to 4 bar).

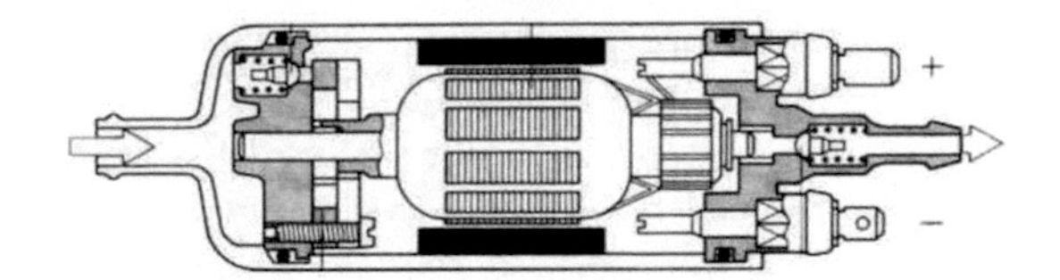

Figure 6.54 Roller cell pump

Roller cell pump The rollers in the roller cell pump are thrown out by centrifugal force when the motor armature and pump rotor spindle rotate. The rotor is fitted eccentrically to the pump body and as the rollers seal against the outer circumference, they create chambers that increase in volume to draw fuel in. They then carry the fuel around and finally discharge it as the chamber volume decreases.

Pump electrical supply The fuel pump electrical supply is live only when the engine is being cranked for starting or is running. The fuel pump electric feed is from a relay that is switched on with the ignition. Safety features are built into the electric control feed to the relay so that it operates only to initially prime the system or when the engine is running. The control functions of the fuel pump relay are usually provided by the fuel control module.

Inertia switch A further safety feature is the use of an inertia switch in the feed from the relay to the fuel pump. This operates, in the event of an accident, to cut the electric feed to the fuel pump and to stop the fuel supply. It is an impact operated switch with a weight that is thrown aside to break the switch contacts. Once the switch has been operated it has to be manually reset.

Fuel filters The fuel filter is an in-line paper element type that is replaced at scheduled service intervals. The filter uses micro-porous paper that is directional for filtration. Filters are marked for fuel flow with an arrow on the casing and correct fitting is essential.

Figure 6.55 An in-line paper element type

Remember! There is a multimedia version of this textbook that includes additional images and interactive features: www.automotivett.org

Fuel pressure regulation The fuel pressure regulator is fitted to maintain a precise pressure at the fuel injector valve nozzles. On port fuel injection systems, a fuel rail is used to hold the pressure regulator and the fuel feed to the injector valves. The injector valves usually fit directly onto or into the fuel rail. The fuel rail holds sufficient fuel to dampen fuel pressure fluctuations and keep the pressure applied at all injector nozzles at a similar level.

Figure 6.56 Fuel pressure regulator on rail

Figure 6.57 Regulator on throttle body

Injector valves The injector valves spray finely atomised fuel into the throttle body or inlet ports depending on the system. The electromagnetic injection valves are actuated by signals from the ECU. The signals are of a precise duration depending on operating conditions but within the range of about 1.5 to 10 milliseconds. This open phase of the injector valve is known as the 'injector pulse width'.

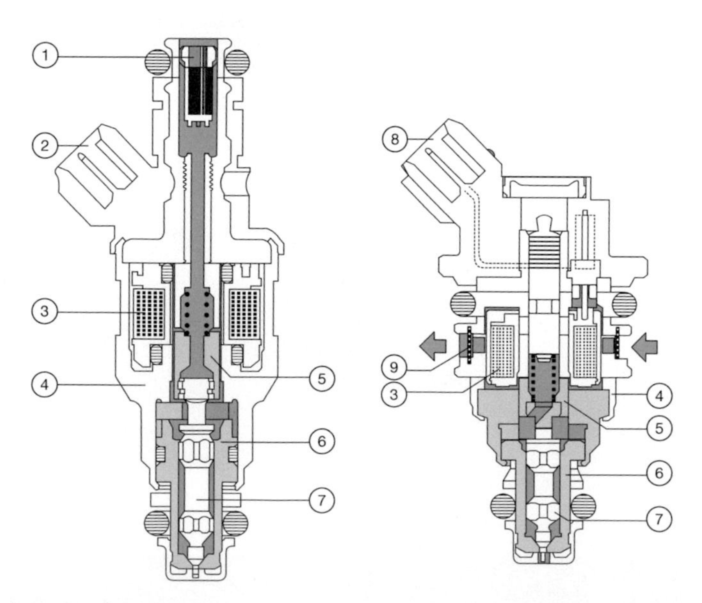

Figure 6.58 Injector features

Solenoid injectors There is a range of individual injector valve designs but all have the same common features. These are an electromagnetic solenoid, with a spring-loaded plunger, connected to a jet needle in the injector valve nozzle. The electrical supply to the solenoid is made from the system relay or ECU. Earthing or grounding the other connection energises the solenoid. This lifts the plunger and jet needle so that fuel is injected for the duration that the electric current remains live. As soon as the electrical supply is switched off in the ECU, a compression spring in the injector valve acts on the solenoid plunger to close the nozzle.

Emission control The ECU also operates the emission control components at appropriate times depending on the engine operating conditions. Typical emission control actuators are the canister purge solenoid valve, the exhaust gas recirculation (EGR) valve and the secondary air solenoid valve. Secondary air is provided by either the air injection reactive or pulse air systems.

Electrical harness The electrical harness for the engine management system is a complex set of cables and sockets. Cables have colour and/or numerical coding and the sockets are keyed so that they can be connected in one way only. Special low resistance connectors are used for low current sensor wiring. Follow manufacturer's data sheets for further technical detail.

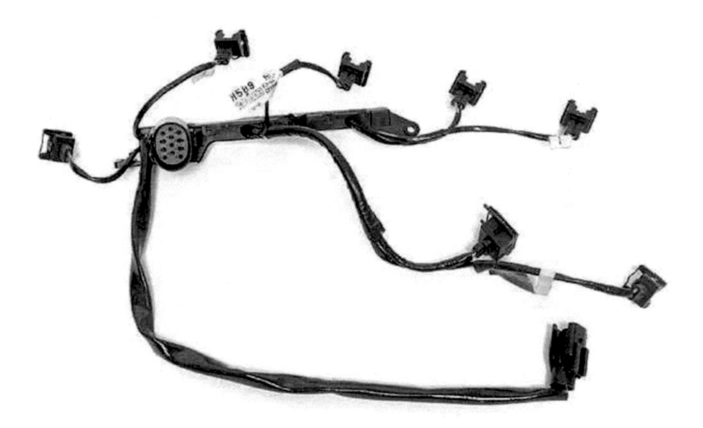

Figure 6.59 Injection wiring harness

Sensors Sensors provide data to the ECU. The engine speed and load conditions are used to calculate the base time value (in milliseconds), for the injector pulse width. A range of correction factors are added to or subtracted from the base time value to suit the engine operating conditions occurring at all instances of time.

Calculation of injection time

Engine speed and position On early electronic fuel injection systems, the engine speed was provided from signals obtained from the ignition low tension primary circuit. On engine management systems, the engine speed and position is required for the ignition and fuel systems.

Figure 6.60 Inductive speed sensor

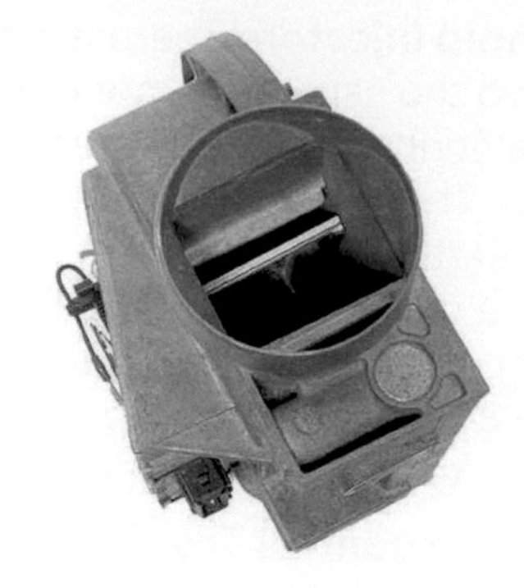

Figure 6.61 Vane type air flow meter

Air flow meter The fuel requirement is calculated in the ECU from the engine speed and load conditions. An air flow meter is one method of measuring the engine load conditions. A variable voltage, corresponding to the measured value at the air flow meter, is used by the ECU to calculate the amount of fuel needed to give a correct air/fuel ratio.

Engine load Engine load can also be determined from the inlet manifold absolute pressure (MAP) and this is used on some systems to provide the data. In these systems, an air flow meter is not used.

Air flow metering There are two main types of air flow meter. These are the vane type (VAF) and the resistive types (MAF). The vane type air flow meter consists of an air passage and damping chamber into which is fitted a fixed pair of flaps (or vanes), which rotate on a spring-loaded spindle. The spindle connects to and operates a potentiometer and switches.

Flap type air flow meter Air flow through the meter acts on the intake air flap to move it in opposition to the spring force. The integral damper flap moves into the sealed damper chamber to smooth out the intake pulses. The degree of flap movement and spindle rotation is measurable at the potentiometer as a variable voltage dependent on position. The voltage signal, together with other signals, is used in the ECU to calculate the fuel requirement.

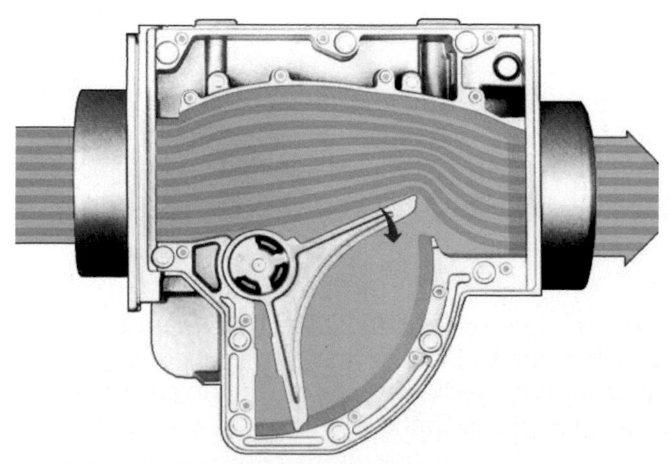

Figure 6.62 Action of the vane air flow meter

Bypass air duct A bypass air duct is built into the housing. This provides for starting without opening the throttle, a smooth air flow during engine idle and a means to adjust the idle mixture.

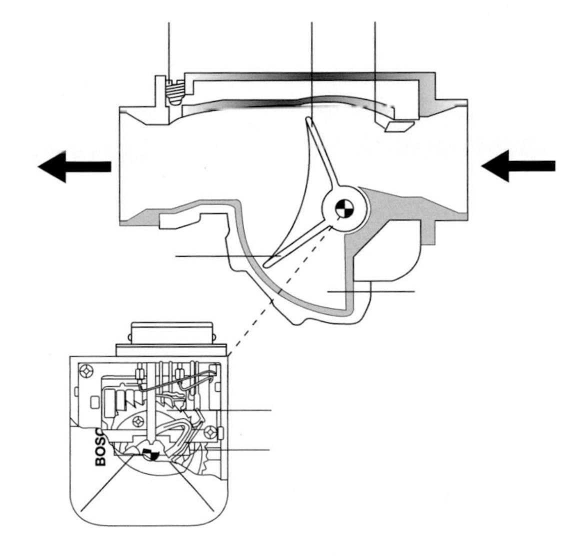

Figure 6.63 Air bypass

Mass air flow meters Mass air flow meters are fitted with two similar resistors inside an air tube. A measurement resistor is heated and often referred to as a hot wire. The other resistor is not heated. It provides a reference value for use in the calculation of the air mass. The control circuit maintains the temperature differential between the two resistors. The signal sent to the ECU is proportional to the current required to heat the measurement resistor and maintain the temperature differential. The output signal from some mass air flow meters is similar to the air vane types. However, some produce a digital output signal.

Manifold absolute pressure sensor On some EFI systems manifold absolute pressure (MAP) sensor signals are used by the ECU to calculate the fuel requirements. These systems do not have an air flow meter. The signals from manifold absolute pressure, engine speed, air charge temperature and throttle position, are compared in the ECU to calculate the injector pulse width.

Figure 6.64 Hot wire/film air flow meter

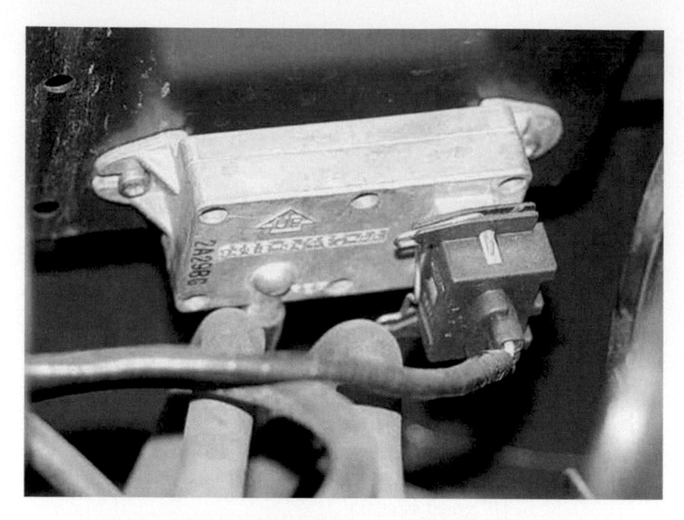

Figure 6.65 MAP sensor

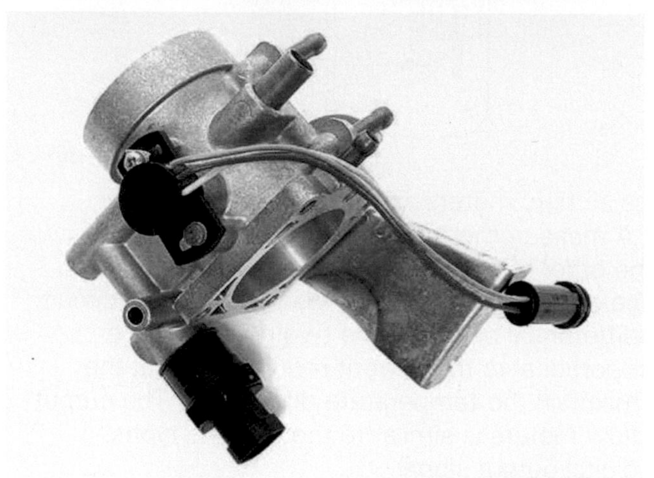

Figure 6.66 Throttle potentiometer

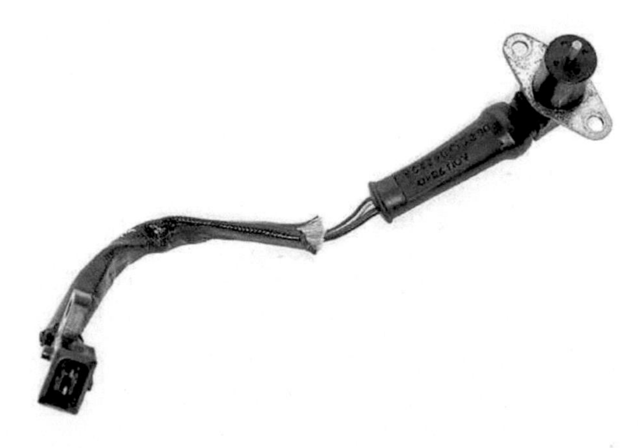

Figure 6.67 Crank sensor

Engine coolant temperature sensor The engine coolant temperature sensor is a negative temperature coefficient (NTC) thermistor. It is of a similar type to the air temperature sensor. It is fitted into the water jacket close to the thermostat or bypass coolant circuit passages. The sensor measures the engine coolant temperature and provides a signal voltage to the ECU. This is

Figure 6.68 Air temperature sensor

Figure 6.69 Coolant thermistor

used for cold start and warm up enrichment as well as fast idle speed control through the idle speed control valve.

Exhaust gas oxygen sensor The Greek letter (l) lambda is used as the symbol for a chemically correct air to fuel ratio. This is the stoichiometric ratio of 14.7 parts of air to 1 part of fuel by mass. Hence, the use of this letter for naming the sensor that is used to control the amount of fuel delivered, so that a very close tolerance to the stoichiometric ratio is maintained.

Lambda sensor The lambda sensor is often known as an exhaust gas oxygen sensor. Some of these sensors are electrically heated. Preheating allows the sensor to be fitted lower down in the exhaust stream and prolongs the life of the active element. The sensor measures the presence of oxygen in the exhaust gas and sends a voltage signal to the engine electronic control unit.

Malfunction indicator light Service and on board diagnostic (OBD) plugs are used for diagnostic and corrective actions with scan tools, dedicated test equipment and other test equipment. If faults are detected the system malfunction indicator lamp on the vehicle fascia will come on. Alternatively, it will fail to go out after the pre-set time duration after switching on the engine. All faults should be investigated as soon as possible. Many electronic systems have a limp home or limited operation strategy program, which allows the vehicle to be driven to a workshop for repair.

Figure 6.70 An exhaust gas oxygen sensor

Look back over the previous section and write out a list of the key bullet points.

Make a simple sketch to show how one of the main components or systems in this section operates.

6.4 Introduction to compression ignition fuel systems

Together with the multimedia resources, this section is ideal material for students working towards the IMI Awards unit L120, the City and Guilds 3902–003 unit and all other similar foundation or introduction level qualifications.

After successful completion of this section you will be able to show you have achieved these objectives:

- Be able to work safely.
- Know the components of compression ignition fuel systems.
- Be able to change fuel and air filters.
- Be aware of environmental considerations.

6.4.1 What you must know about compression ignition fuel systems

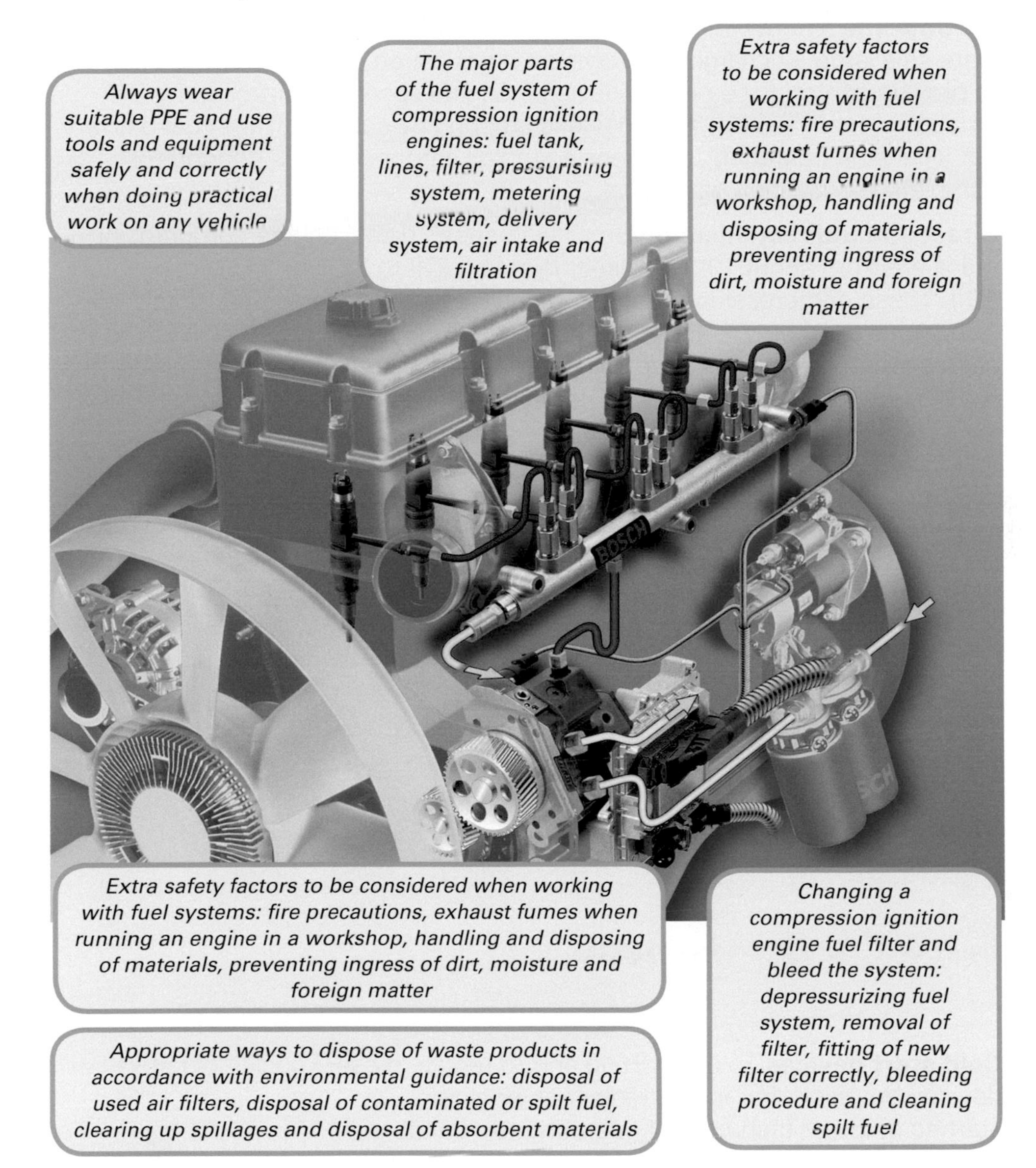

6.4.2 Diesel injection

Diesel fuel injection systems Diesel engines have the fuel injected into the combustion chamber where it is ignited by heat in the air charge. This is known as compression ignition (CI) because no spark is required. The high temperature needed to ignite the fuel is obtained by a high compression of the air charge.

High pressure pump Diesel fuel is injected under high pressure from an injector nozzle, into the combustion chambers. The fuel is pressurised in a diesel injection pump. It is supplied and distributed to the injectors through high pressure fuel pipes. Some engines use a unit injector where the pump and injector are combined in a single unit. The high pressure generation is from a direct acting cam or a separate pump.

Air flow The air flow into a diesel engine is usually unobstructed by a throttle plate so a large air charge is always provided. Throttle plates may be used to provide control for emission devices. Engine speed is controlled by the amount of fuel injected. The engine is stopped by cutting off the fuel delivery. For all engine operating conditions a surplus amount of air is needed for complete combustion of the fuel.

Direct and indirect injection Small high-speed diesel engine compression ratios are from about 19:1 for direct injection (DI) to 24:1 for indirect injection (IDI). These compression ratios are capable of raising the air charge to temperatures of between 500°C and 800°C. Very rapid combustion of the fuel occurs when it is injected into the hot air charge.

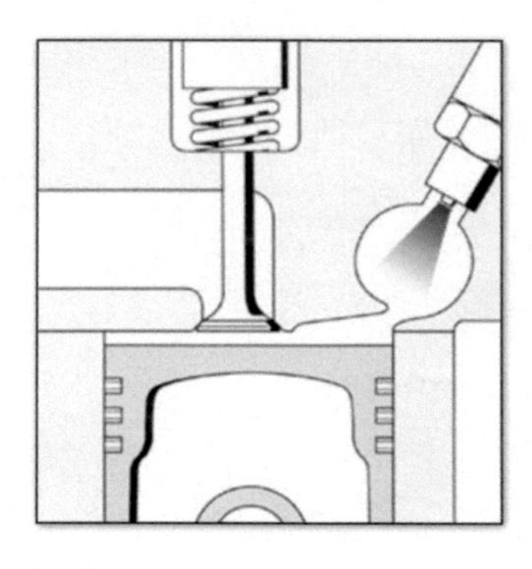

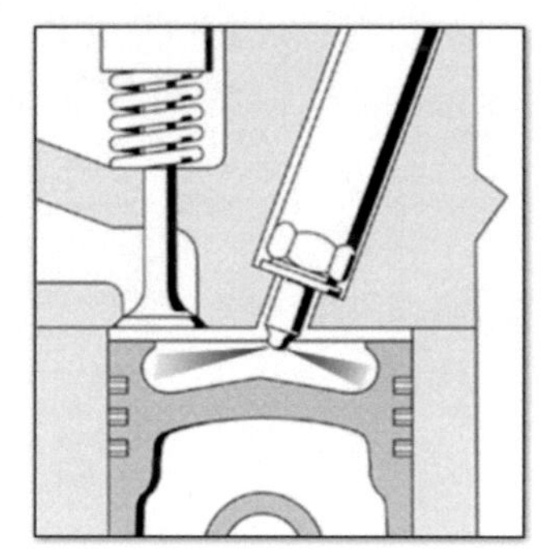

Figure 6.71 Indirect and direct injection

Cold start devices To aid starting and to reduce diesel knock, cold start devices are used. For indirect injection engines, starting at lower than normal operating temperatures requires additional combustion chamber heating. For direct injection engines, cold start devices are only required in frosty weather.

Particulates Another exhaust gas constituent is particulate emissions. These result from incomplete combustion of the fuel. Particulates are seen as black carbon smoke in the exhaust under heavy load or when fuel delivery and/or timing is incorrect. White smoke may also be visible at other times, such as when the injection pump timing is incorrect. It also occurs when compression pressures are low or when coolant has leaked into the combustion chambers.

Figure 6.72 Glow plug

Direct and indirect injection Direct injection (DI) is made into a combustion chamber formed in the piston crown. Indirect injection (IDI) is made into a pre-combustion chamber in the cylinder head. Direct injection engines are generally more efficient but the indirect types are quieter in operation. The internal stresses in the engine are very high. Direct injection produces a higher detonation stress than indirect injection and therefore the smaller engines tended, until recently, to be the indirect type.

Electronic control Recent developments in electronic diesel fuel injection control have made it possible to produce small direct injection engines. It is probable that all new designs of diesel engine will be of this type. Diesel engines are built to withstand the internal stresses, which are greater than other engines. Diesel engines are particularly suitable for turbocharging. This improves power and torque outputs.

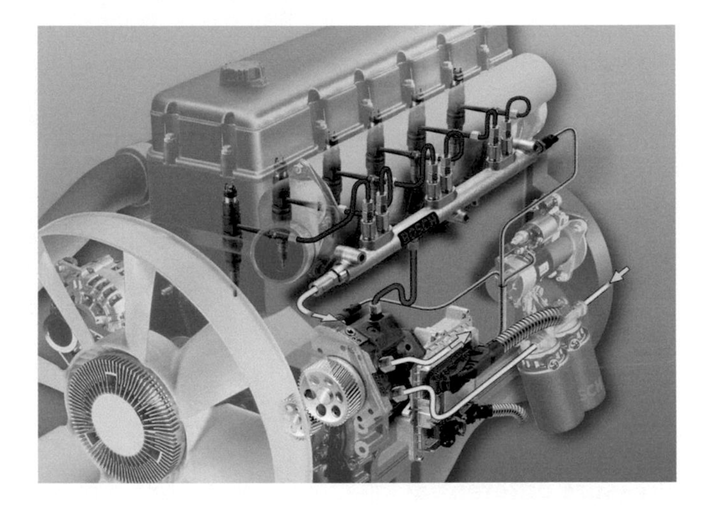

Figure 6.73 Common rail injection

Exhaust gas recirculation Exhaust gas recirculation (EGR) has two advantages for diesel engine operation. EGR is usually used to reduce nitrogen oxide (NOx) emissions and this is true for diesel engines. Additionally, a small quantity of hot exhaust gas in the air charge of a cold engine helps to reduce the delay period and the incidence of cold engine diesel knock.

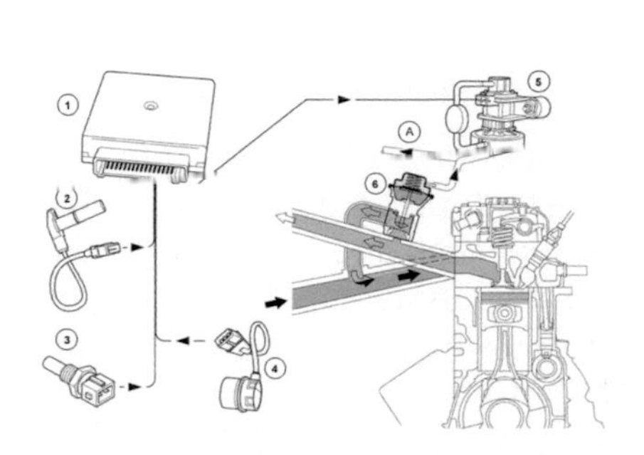

Figure 6.74 EGR

Figure 6.75 Inside a catalytic converter

Catalytic converters Many modern diesel-engined vehicles are fitted with oxidation catalytic converters that work in conjunction with other emission components to reduce hydrocarbon and particulate emissions. Turbo charging, EGR and catalytic converters are described in the Air Supply, Exhaust and Emission Control learning programme.

Injection pressures The fuel systems for direct and indirect injection are similar and vary only in injection pressures and injector types. Until recently, all light high-speed diesel engines used rotary diesel fuel injection pumps. These pumps producing injection pressures of over 100 bar for indirect engines. However, these can rise up to 1000 bar at the pump outlet, for turbocharged direct injection engines.

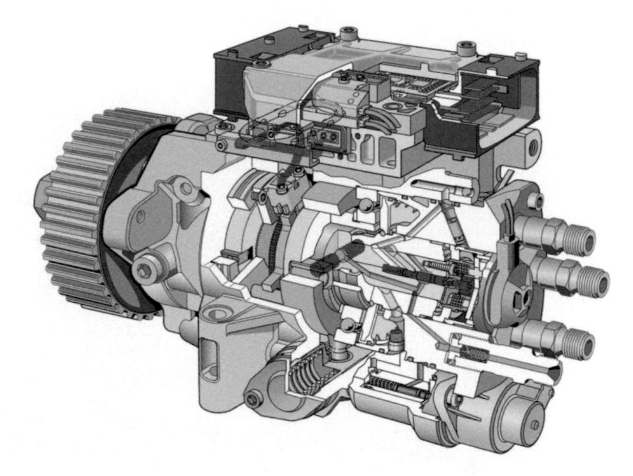

Figure 6.76 Bosch injection pump

Pressure differential Injectors operate with a pulsing action at high pressure to break the fuel down into finely atomised parts. Atomisation is critical to good fuel distribution in the compressed air charge. The air charge pressure may be in excess of 60 bar. The pressure differential, between the fuel injection pressure and air charge pressure, must be sufficient to overcome the resistance during injection. This will also give good fuel atomisation and a shorter injection time.

Swirl An aid to good fuel distribution in the air charge is the swirl in the air flow induced in the inlet manifold. This is created by the combustion chamber design. Air flow into and out of the pre-combustion chamber produces a swirl in the chamber. These chambers are often referred to as swirl chambers. The 'bowl in piston' combustion chambers, of direct injection engines, are shaped to maintain the induction air swirl during compression and combustion.

Diesel fuel injection components The main components of a diesel fuel system provide for either the low pressure or the high pressure functions. The low pressure components are the fuel tank, the fuel feed and return pipes and hoses, a renewable fuel filter with a water trap and drain tap, and a priming or lift pump. Fuel heaters may be fitted in the filter housing to reduce the risk of paraffin separation and waxing at freezing temperatures.

High pressure components The high pressure components are the fuel injector pump, the high pressure pipes and the injectors. Other components provide for cold engine starting. Electronically controlled systems include sensors, an electronic diesel control (EDC) module and actuators in the injection pump.

Low pressure components The fuel tank is a pressed steel sealed unit, treated both inside and out with anti-corrosion paint. The inside is treated in order to resist corrosion from water that accumulates at the bottom of the tank. Some modern tanks are manufactured from a plastic compound that is burst proof in an accident. They are also unaffected by the diesel fuel, which can attack some plastic materials.

Low pressure fuel lines Low pressure fuel lines are steel or hard plastic and connections made with short hoses clamped at each end. New vehicles are using quick coupling connections for ease of service and assembly operations. The feed lines run from the tank to the filter and then onto the injection pump. A low pressure return line is used to maintain a fuel flow through the injection pump and the fuel injectors for lubrication and cooling. The return carries fuel back to the filter housing or the fuel tank.

Fuel filter The fuel filter is a micro-porous paper element in a replaceable canister or detached filter bowl. The filter includes a water and sediment trap and tap for draining the water. Many vehicles have a sensor in the water trap. This completes a warning lamp circuit when water is detected above a certain level. All diesel fuel entering the injection pump and injectors must be fully

Figure 6.77 Filter and housing

filtered. The internal components of the pump and injectors are manufactured to very fine tolerances. Even very small particles of dirt could be damaging to these components.

Fuel heating Fuel heating may be provided from the engine coolant or by an electric heater element in the filter housing The fuel is lifted from the fuel tank to the injection pump by the transfer pump in the injection pump on some vehicles. This is possible where the distance and height of lift are of small dimensions.

Fuel lift pump For improved delivery and for priming the injection pump another pump may be necessary. A conventional fuel lift pump driven from

Remember! There is a multimedia version of this textbook that includes additional images and interactive features: www.automotivett.org

the engine camshaft is a common method. These pumps in some instances have an external operating lever. Hand operated priming pumps are fitted for use when the vehicle runs out of fuel. They are also used for service operations such as when the filter is changed. Many modern injector pumps are self-priming.

Figure 6.78 Heated fuel filter housing

Injection pump The injector pump shown is a rotary distributor type pump. These pumps are filled with diesel fuel, which provides not only fuel for the engine, but also for full lubrication and cooling of the pump. These pumps are made from specially manufactured materials with surface treatments. Parts are lapped together to give very fine tolerances. Only clean and filtered fuel should be used to avoid damage to these parts.

Figure 6.79 Fuel priming pump

Types of pump There are two types of pump with different internal operation. Two major original designers make or license the manufacture of these pumps. The Lucas DP and Bosch VR pumps are radial-piston designs. They use opposing pistons or plungers inside a cam ring to produce the high pressure. Bosch V series pumps are axial-piston designs having a roller ring and cam plate attached to an axial piston or plunger in the distributor head to generate the high pressure.

Fuel injectors The fuel injectors are fitted into the cylinder head with the nozzle tip projecting into the pre-combustion (IDI) or combustion chamber (DI). The injectors for indirect combustion are of a pintle or 'pintaux' design and produce a conical spray pattern on injection. The injectors for direct injection (DI) are of a pencil-type multi-hole design that produces a broad distribution of fuel on injection.

Figure 6.80 Bosch pumps

Figure 6.81 DI injector

Injector operation Fuel injectors are held closed by a compression spring. They are opened by hydraulic pressure when it is sufficient to overcome the spring force on the injector needle. The hydraulic pressure is applied to a face on the needle where it sits in a pressure chamber. The fuel pressure needed is in excess of 100 bar (1500 psi). This pressure lifts the needle and opens the nozzle, so that fuel is injected in a fine spray pattern into the combustion chamber.

Figure 6.82 IDI injector

Injector spray The pressure drops when fuel is injected and the spring force on the needle closes the injector. This is immediately followed up by a build-up of pressure that again opens the nozzle. This results in a cycle of oscillations of the needle to give a finely atomised and almost continuous spray. The spray continues until the pump pressure is reduced at the end of the delivery stroke.

Look back over the previous section and write out a list of the key bullet points.

Construct a word search grid using some key words from this section. About 10 words in a 12x12 grid is usually enough.

6.4.3 Exhaust systems

System requirements The exhaust system has to carry the exhaust gases out of the engine to a safe position on the vehicle, silence the exhaust sound and cool the exhaust gases. It also has to match the engine gas flow, resist internal corrosion from the exhaust gas and resist external corrosion from water and road salt.

System components The exhaust system consists of the exhaust manifold, silencers, mufflers, expansion boxes and resonators. It also has down or front pipes, intermediate and tail pipes, heat shields and mountings. Also included are one or more catalytic converters, one or two lambda sensors and an outlet for the exhaust gas recirculation system.

Figure 6.83 Exhaust system

High temperatures The exhaust gases are at a very high temperature when they leave the combustion chambers and pass through the exhaust ports. The exhaust manifold is made from cast iron in order to cope with the high temperatures. The remainder of the exhaust system is made from steel, which is alloyed and treated to resist corrosion.

Down pipe or front pipe The down pipe, or front pipe, is attached to the manifold with a flat, or ball, flange. This joint is subject to bending stresses with the movement of the engine in the vehicle. To accommodate the movement and reduce stress fractures, many flange connections have a flexible coupling made from a ball flange joint and compression springs on the mounting studs.

Exhaust movement Another system to accommodate movement is a flexible pipe constructed from interlocking stainless-steel coils, or rings. Where a flexible joint is not required, the front pipe may be supported by a bracket welded to the pipe which is bolted to a convenient position on the engine or gearbox. Where a catalytic converter is used, it is fitted to the front pipe so that the exhaust heat is used to aid the chemical reactions taking place within the catalytic converter. The front pipe connects to an expansion box or silencer. The exhaust gases are allowed to expand into this box and begin to cool. They contract on cooling and slowdown in speed.

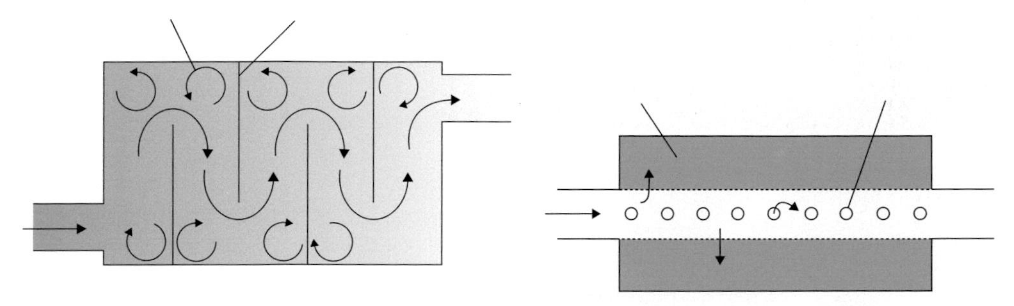

Figure 6.84 Baffle silencer

Figure 6.85 Absorption silencer

Silencers or mufflers Silencers are constructed as single- or twin-skin boxes, and there are two main types: The absorption type, which uses glass fibre or steel wool to absorb the sound; and the baffle type, which uses a series of baffles to create chambers. In the baffle type, the exhaust gases are transferred from a perforated inlet pipe to a similarly perforated outlet pipe. These silencers have a large external surface area so that heat is radiated to the atmosphere. Additional pipes and silencers carry the exhaust gas to the rear.

Joints Pipes are joined together by a flange, or clamp, fitting. Flange connections have a heat-resistant gasket and through-bolts to hold the flange together. Clamp fittings are used where pipes fit into each other. The larger pipe is toward the front and the smaller pipe fits inside. A ring clamp, or 'U' bolt and saddle, are tightened around the pipes to give a gas-tight seal. An exhaust paste is usually applied to improve the seal of the joint. The exhaust system must be sealed to prevent toxic exhaust gases from entering the passenger compartment

Exhaust mountings A small water-drain hole may be used on the underside of some silencers. This is to reduce internal corrosion from standing water in the silencer body forming on short-journey usage. The exhaust is held underneath the vehicle body on flexible mountings. These are usually made from a rubber compound and many are formed as a large ring that fits on hooks on the vehicle and the exhaust-pipe brackets. Other mountings are bonded-rubber blocks on two steel plates.

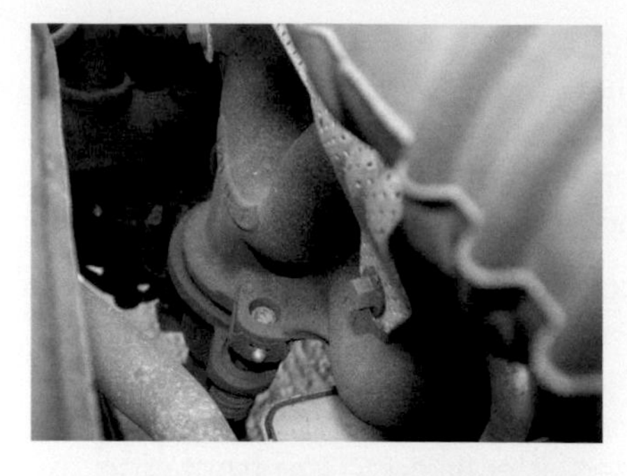

Figure 6.86 Flange connections

Figure 6.87 Pipe connections

Figure 6.88 Flexible . . .

Figure 6.89 Mountings

Look back over the previous section and write out a list of the key bullet points.

Make a simple sketch to show how one of the main components or systems in this section operates.

6.5 Introduction to Vehicle Ignition Systems

Together with the multimedia resources, this section is ideal material for students working towards the IMI Awards unit L112, the City and Guilds 3902–003 unit and all other similar foundation or introduction level qualifications.

After successful completion of this section you will be able to show you have achieved these objectives:

- Be able to work safely.
- Know the main components of a vehicle ignition system.
- Know how to replace a vehicle ignition component.
- Be aware of environmental considerations.

6.5.1 What you must know about ignition systems

6.5.2 Ignition overview

Purpose The purpose of the ignition system is to supply a spark inside the cylinder, near the end of the compression stroke, to ignite the compressed charge of air/fuel vapour. For a spark to jump across an air gap of 1.0 mm under normal atmospheric conditions (1 bar) a voltage of 4 to 5 kV is required. For a spark to jump across a similar gap in an engine cylinder, having a compression ratio of 8:1 approximately 10 kV is required. For higher compression ratios and weaker mixtures, a voltage up to 20 kV may be necessary. The ignition system has to transform the normal battery voltage of 12 V to approximately 8 to 20 kV and, in addition, has to deliver this high voltage to the right cylinder, at the right time. Some ignition systems will supply up to 40 kV to the spark plugs.

Figure 6.90 Combustion taking place (Source: Ford Media)

Conventional ignition is the forerunner of the more advanced systems controlled by electronics. It is worth mentioning at this stage, however, that the fundamental operation of most ignition systems is very similar. One winding of a coil is switched on and off causing a high voltage to be induced in a second winding. The basic types of ignition system can be classified as shown in the table.

Type	Conventional	Electronic	Programmed	Distributorless
Trigger	Mechanical	Electronic	Electronic	Electronic
Advance	Mechanical	Mechanical	Electronic	Electronic
Voltage source	Inductive	Inductive	Inductive	Inductive
Distribution	Mechanical	Mechanical	Mechanical	Electronic

Engine management Modern ignition systems now are part of the engine management, which controls fuel delivery, ignition and other vehicle functions. These systems are under continuous development and reference to the manufacturer's workshop manual is essential when working on any vehicle. The main ignition components are the engine speed and load sensors, knock sensor, temperature sensor and the ignition coil. The ECU reads from the sensors, interprets and compares the data, and sends output signals to the actuators. The output component for ignition is the coil.

Figure 6.91 Electronic ignition and injection

Developments Ignition systems continue to develop and will continue to improve. However, keep in mind that the simple purpose of an ignition system is to ignite the fuel air mixture every time at the right time. And, no matter how complex the electronics may seem, the high voltage is produced by switching a coil on and off.

Generation of high voltage If two coils (known as the primary and secondary) are wound on to the same iron core then any change in magnetism of one coil will induce a voltage in to the other (see electrical chapter for more details). This happens when a current is switched on and off to the primary coil. If the number of turns of wire on the secondary coil is more than the primary a higher voltage can be produced. This is called transformer action and is the principle of the ignition coil.

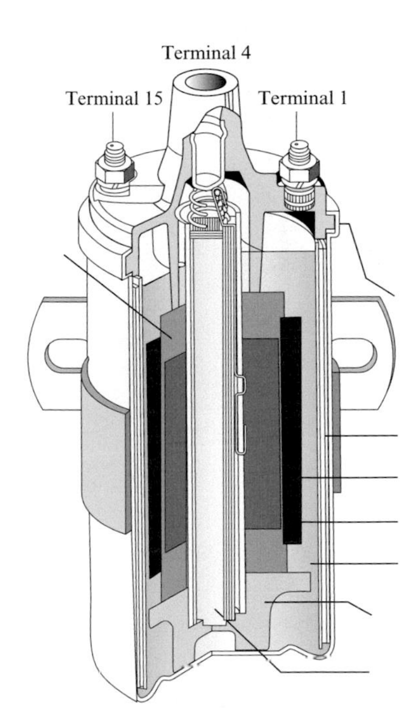

Figure 6.92 Traditional ignition coil

Value of this 'mutually induced' voltage depends upon:

- primary current
- turns ratio between primary and secondary coils
- the speed at which the magnetism changes.

The two windings are wound on a laminated iron core to concentrate the magnetism. This is how all types of ignition coil are constructed.

Ignition timing For optimum efficiency the ignition timing (or advance angle) should be such as to cause the maximum combustion pressure to occur about 10o after TDC. The ideal ignition timing is dependent on two main factors, engine speed and engine load. An increase in engine speed requires the ignition timing to be advanced. The cylinder charge, of air fuel mixture, requires a certain time to burn (something like 2ms). At higher engine speeds the time taken for the piston to travel the same distance reduces. Advancing the time of the spark ensures full burning is achieved.

Engine load A change in timing due to engine load is also required as the weaker mixture used on low load conditions burns at a slower rate. In this situation further ignition advance is necessary. Greater load on the engine requires a richer mixture, which burns more rapidly. In this case some retardation of timing is necessary. Overall, under any condition of engine speed and load an ideal advance angle is required to ensure maximum pressure is achieved in the cylinder just after top dead centre. The ideal advance angle may also be determined by engine temperature and any risk of detonation.

Spark advance is achieved in a number of ways. The simplest of these being the mechanical system comprising of a centrifugal advance mechanism and a vacuum (load sensitive) control unit. Manifold depression is almost inversely proportional to the engine load. However, I prefer to consider manifold pressure instead of vacuum or depression even though it is lower than atmospheric pressure. The manifold absolute pressure (MAP) is therefore proportional to engine load. Digital ignition systems adjust the timing in relation to the temperature as well as speed and load. The values of all ignition timing functions are combined either mechanically or electronically in order to determine the ideal ignition point.

The **energy storage** takes place in the ignition coil. The energy is stored in the form of a magnetic field. To ensure the coil is charged before the ignition point a dwell period is required. Ignition timing is at the end of the dwell period as the coil is switched off.

Early ignition system Very early cars used something called a magneto, which is a story for another time, but here is a nice picture of one anyway!

Figure 6.93 First Bosch high-voltage magneto ignition system with spark plug in 1902 (Source: Bosch Media)

Figure 6.94 Contact breaker system

Mechanical switching For many years ignition systems were mechanically switched and distributed. The table gives an overview of the components on this earlier system.

Modern systems All current vehicle ignition systems are electronically switched and most are now digitally controlled as part of the engine management system. However, there are many vehicles out there still using conventional electronic ignition so the next main section will give an overview of these systems.

Spark plug	Seals electrodes for the spark to jump across in the cylinder. Must withstand very high voltages, pressures and temperatures
Ignition coil	Stores energy in the form of magnetism and delivers it to the distributor via the HT lead. Consists of primary and secondary windings
Ignition switch	Provides driver control of the ignition system and is usually also used to cause the starter to crank
Contact breakers (breaker points)	Switches the primary ignition circuit on and off to charge and discharge the coil. The contacts are operated by a rotating cam in the distributor
Capacitor (condenser)	Suppresses most of the arcing as the contact breakers open. This allows for a more rapid break of primary current and hence a more rapid collapse of coil magnetism which produces a higher voltage output
Distributor	Directs the spark from the coil to each cylinder in a pre-set sequence
Plug leads	Thickly insulated wires to connect the spark from the distributor to the plugs
Centrifugal advance	Changes the ignition timing with engine speed. As speed increases the timing is advanced
Vacuum advance	Changes timing depending on engine load. On conventional systems the vacuum advance is most important during cruise conditions

Figure 6.95 Traditional ignition coil

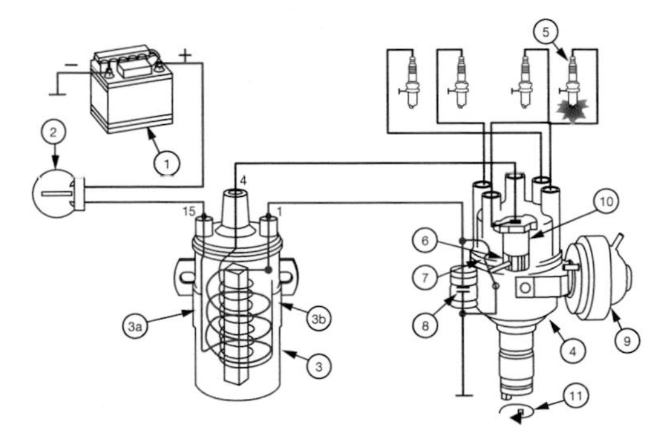

Figure 6.96 Ignition circuit of an early system: 1 – battery, 2 – ignition key switch, 3 – coil, a –primary, b – secondary winding, 4 – distributor body containing centrifugal (speed) advance/retard mechanism, 5 – spark plugs, 6 – cam (with a lobe for each cylinder), 7 – contact breakers (points), 8 –condensor (capacitor), 9 – vacuum (load) advance/ retard mechanism

Figure 6.97 Traditional system using a distributor

Look back over the previous section and write out a list of the key bullet points.

Construct a crossword puzzle using important words from this section. Hint: use the ATT glossary where you can copy the words and definitions (clues!). About 20 words is a good puzzle.

6.5.3 Spark plugs and leads

Overview The simple requirement of a spark plug is that it must allow a spark to form within the combustion chamber, to initiate combustion. In order to do this the plug has to withstand a number of severe conditions. It must withstand severe vibration and a harsh chemical environment. Finally, but perhaps most important, the insulation properties must withstand voltages pressures up to 40 kV.

Standard spark plug The centre electrode is connected to the top terminal by a stud. The electrode is constructed of a nickel-based alloy. Silver and platinum are also used for some applications. If a copper core is used in the electrode this improves the thermal conduction properties. The insulating material is ceramic based and of a very high grade. Flash over or tracking down the outside of the plug insulation is prevented by ribs which effectively increase the surface distance from the terminal to the metal fixing bolt, which is of course earthed to the engine.

Temperature Due to many and varied constructional features involved in the design of an engine, the range of temperatures a spark plug is exposed to can vary significantly. The operating temperature of the centre electrode of a spark plug is critical. If the temperature becomes too high then pre-ignition may occur where the fuel-air mixture may be ignited due to the incandescence of the plug electrode. If the electrode temperature is too low, then carbon and oil fouling can occur as deposits are not burnt off. The ideal operating temperature of the plug electrode is between 400 and 900°C.

The **heat range** of a spark plug is a measure of its ability to transfer heat away from the centre electrode. A hot running engine will require plugs with a higher

Figure 6.98 Modern high performance spark plug

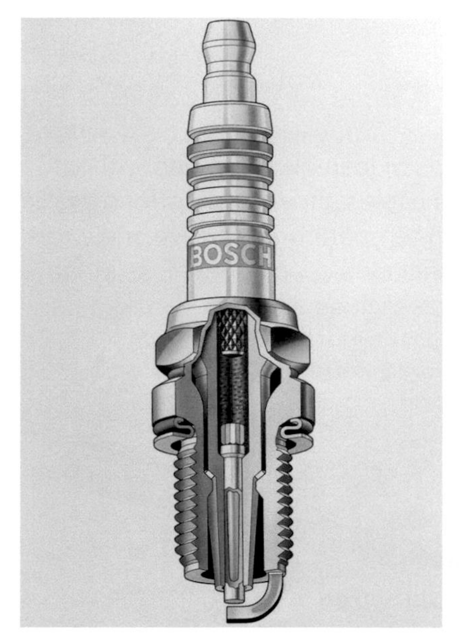

Figure 6.99 Cutaway section of a spark plug

thermal ability than a colder running engine. Note that hot and cold running of an engine in this sense refers to the combustion temperature, not to the cooling system.

Spark plug **electrode gaps** in general have increased as the power of the ignition systems driving the spark has increased. The simple relationship between plug gap and voltage required is that as the gap increases so must the voltage (leaving aside engine operating conditions). Further, the energy available to form a spark at a fixed engine speed is constant, which means that a larger gap using higher voltage will result in a shorter duration spark. A smaller gap will allow a longer duration spark. For cold starting an engine and for igniting weak mixtures the duration of the spark is critical. Likewise the plug gap must be as large as possible to allow easy access for the mixture to prevent quenching of the flame. The final choice is therefore a compromise reached through testing and development of a particular application. Plug gaps in the region of 0.6 to 1.2mm seem to be the norm at present.

High tension (HT) is just an old fashioned way of saying high voltage. HT components such as plug leads, must meet or exceed stringent ignition product requirements, such as:

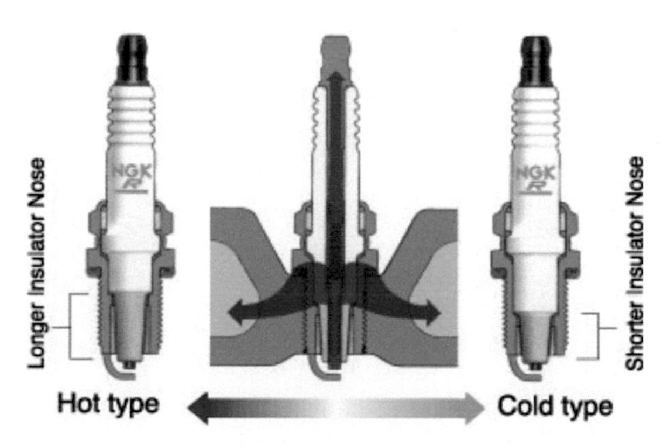

Figure 6.100 Heat-loss paths: 1 – cold plug, 2 – hot plug, 3 – temperature (the cold plug is able to transfer heat more easily so is suitable for a hot engine)

Remember! There is a multimedia version of this textbook that includes additional images and interactive features: www.automotivett.org

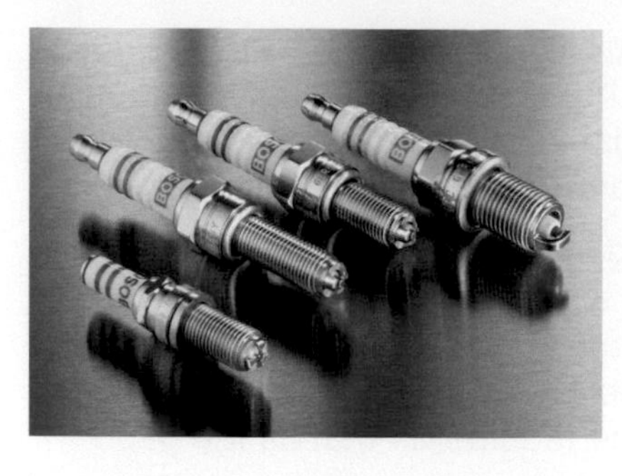

Figure 6.101 A range of spark plugs (Source: Bosch Media)

- Insulation to withstand 50 000 V.
- Temperatures from 40°C to 260°C (40°F to 500°F).
- Radio frequency interference suppression.
- 160 000 km (100 000 mile) product life.
- Resistance to ozone, corona and fluids.
- 10-year durability.

HT cables must meet the increased energy needs of lean-burn engines without emitting electromagnetic interference (EMI). The cables shown here offer metallic and non-metallic cores, including composite, high-temperature resistive and wire-wound inductive cores. Conductor construction includes copper, stainless steel, Delcore, CHT and wire-wound. Jacketing materials include silicone.

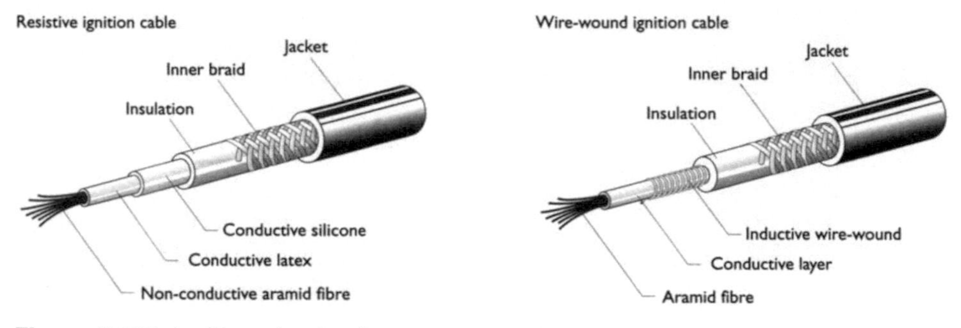

Figure 6.102 Ignition plug leads

Look back over the previous section and write out a list of the key bullet points.

Create a mind map to illustrate the important features of a component or system in this section.

Now complete the multiple choice quiz associated with this topic/subject area.

6.6 Engine systems puzzles

6.6.1 Cryptograms

A	B	C	D	E	F	G	H	I	J	K	L	M	N	O	P	Q	R	S	T	U	V	W	X	Y	Z
	13																	12							

S	_	_	_		S	_	_	_	_	_	_		B	_	_	_		B	_	_	_
12	21	11	18		12	16	21	2	2	24	2		13	5	23	22		13	8	25	19

Cryptogram 6.1 Otto:

6.6.2 Anagrams

Burial Tonic	Brutal Ionic
Ironical But	Curtain Boil

Anagrams 6.1 Keep surfaces apart: what word did I use to get these anagrams?

6.6.3 Word search

```
S S P I U N N T K Y T X T C S
A G L H C P E I A U S C O O U
C C N X O W S S V S X M K R Q
E I T I D W J T G Z B J T M J
X L J N T F L S R U D T V V H
H U Y N Z N I M S O E O G G O    CAMSHAFT
A A Q Q H C U T D T K B Q O B    COMBUSTION
U R O B T O I O E B M E G M R    COMPONENTS
S D C O U O E N M Z R E X E V    CYLINDER
T Y C X N B T K R V J B D A K    EXHAUST
V H A J Z Z H Q G F S N L G W    HYDRAULIC
S T N E N O P M O C I V D T Z    MOUNTINGS
C A M S H A F T H L E B U R Y    PISTON
N O T S I P D S Y S X W V H D    UPSTROKE
L S J C H F U C G T A V U H L    VALVES
```

Word search 6.1 Mechanical

6.6.4 Crosswords

Across

2 Engine position when the piston moves past its highest point (1,1,1,1)

9 Reciprocating component in an engine

10 Moving backwards and forwards

15 Component that allows air or air/fuel mixture into an engine (5,5)

16 Oil, for example, to prevent surface contact

19 Sealing devices usually made of cast iron to make a gas tight seal in a cylinder (6,5)

20 Used to open intake and exhaust valves of an engine

21 The conventional operating principle for the modern engine (4,5,5)

22 The engine component that joins the piston to the crankshaft (10,3)

Down

1 The item that closes and holds closed an inlet or exhaust valve

3 A toothed strip that drives the camshaft (6,4)

4 The reduction in volume of a gas when it is squashed

5 Lower part of an engine or gearbox usually containing lubrication oil

6 A stick used to check fluid level

7 The burning of fuel in air

8 Rotary engine using a three-cornered rotor in a trochoidal chamber

11 The main body of an engine (8,5)

12 Engine where fuel is burnt inside cylinders (8,10)

13 Engine position when the piston is at its lowest point (6,4,5)

Crossword 6.1 Mechanical and lubrication

CHAPTER 7

Electrical systems

7.1 Vehicle lighting system maintenance

Together with the multimedia resources, this section is ideal material for students working towards the IMI Awards unit L114, the City and Guilds 3902–005 unit and all other similar foundation or introduction level qualifications.

After successful completion of this section you will be able to show you have achieved these objectives:

- Be able to work safely.
- Know about vehicle lighting systems components.
- Know how vehicle lighting systems operate.
- Be able to replace lighting system components.

Automotive Technician Training: Entry Level 3. 978-0-415-72040-3.
© Tom Denton. Published by Taylor & Francis. All rights reserved.

7.1.1 What you must know about lighting systems

Always wear suitable PPE and use tools and equipment safely and correctly when doing practical work on any vehicle

The colour of lamps that are legally required: headlamp, sidelamp, indicators, brake lights, rear lamps, reverse lamps, fog lamps, white, amber, red and yellow

Correct method to replace a headlamp unit: checking lamp cool, switched off, correct removal, correct refitting and avoiding damage to vehicle, and the correct method to align a headlamp to within legal requirements: preparation of equipment prior to use, correct alignment of equipment to vehicle, identification of correct adjuster, correct alignment

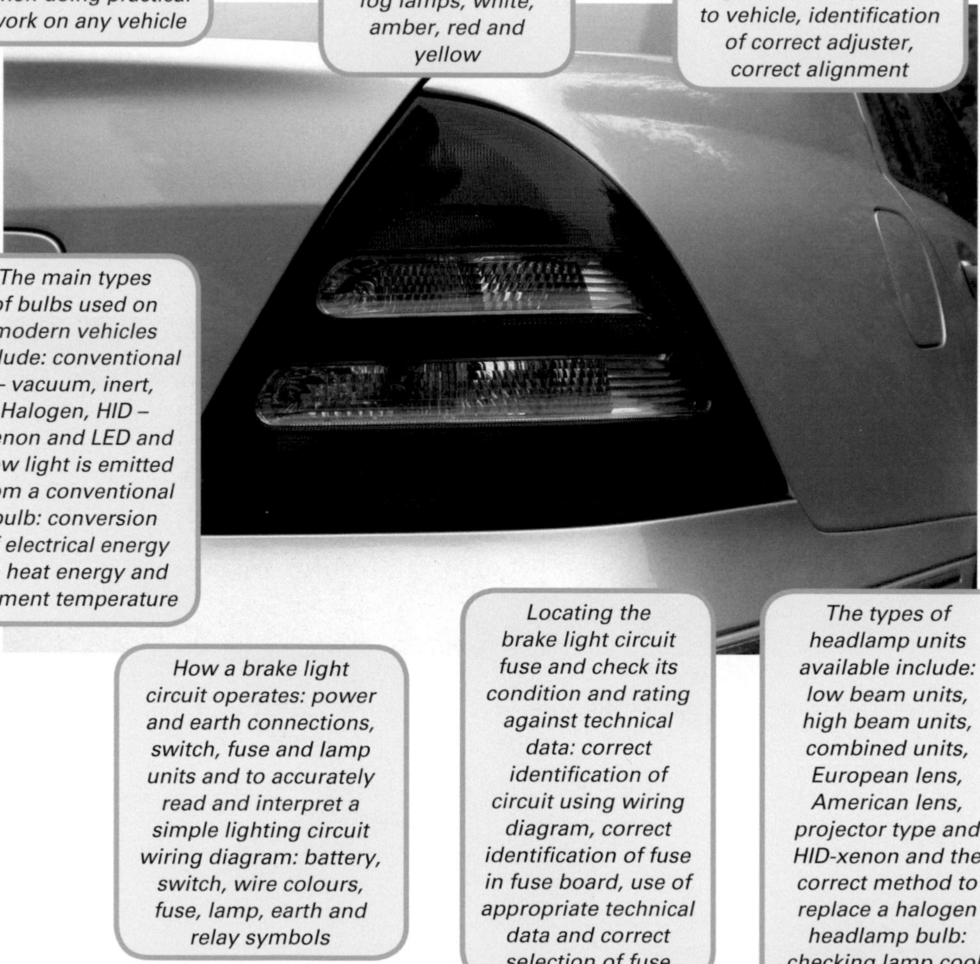

The main types of bulbs used on modern vehicles include: conventional – vacuum, inert, Halogen, HID – Xenon and LED and how light is emitted from a conventional bulb: conversion of electrical energy to heat energy and filament temperature

How a brake light circuit operates: power and earth connections, switch, fuse and lamp units and to accurately read and interpret a simple lighting circuit wiring diagram: battery, switch, wire colours, fuse, lamp, earth and relay symbols

Locating the brake light circuit fuse and check its condition and rating against technical data: correct identification of circuit using wiring diagram, correct identification of fuse in fuse board, use of appropriate technical data and correct selection of fuse

The types of headlamp units available include: low beam units, high beam units, combined units, European lens, American lens, projector type and HID-xenon and the correct method to replace a halogen headlamp bulb: checking lamp cool, check switched off, correct removal, refitting avoiding contact with quartz envelope

7.1.2 Electrical components and circuits

A switch is a simple device used to break a circuit, that is, it prevents the flow of current. A wide range of switches is used. Some switches are simple on/off devices such as an interior light switch on the door pillar. Other types of switch are more complex. They can contain several sets of contacts to control, for example, the indicators, headlights and horn. These are described as multifunction switches.

Relays A relay is a very simple device. It can be thought of as a remote controlled switch. A very small electric current is used to magnetise a small winding. The magnetism then causes some contacts to close, which in turn can control a much heavier current. This allows small delicate switches to be used, to control large current users, such as the headlights or the heated rear window.

Figure 7.1 Simple 'cube' relay

Some form of **circuit protection** is required to protect the electrical wiring of a vehicle and to protect the electrical and electronic components. It is now common practice to protect almost all electric circuits with a fuse. A fuse is the weak link in a circuit. If an overload of current occurs then the fuse will melt and disconnect the circuit before any serious damage is caused. Automobile fuses are available in three types: glass cartridge, ceramic and blade type. The blade type is now the most popular choice due to its simple construction and reliability. Fuses are available in a number of rated values. Only the fuse recommended by the manufacturer should be used.

A **fuse** protect the device as well as the wiring. A good example of this is a fuse in a wiper motor circuit. If a value were used, which is much too high then it would still protect against a severe short circuit. However, if the wiper blades froze to the screen, a large value fuse might not protect the motor from overheating.

Current rating	Colour code
3	Violet
4	Pink
5	Clear/beige
7.5	Brown
10	Red
15	Blue
20	Yellow
25	Neutral/white
30	Green

Fusible links Fusible links in the main output feeds from the battery protect against major short circuits in the event of an accident or error in wiring connections. These links are simply heavy-duty fuses and are rated in values such as 50, 100 or 150A.

Figure 7.2 These links connect to the battery

Circuit breakers Occasionally circuit breakers are used in place of fuses, this being more common on heavy vehicles. A circuit breaker has the same rating and function as a fuse but with the advantage that it can be reset.

Terminals and connectors Many types of terminals are available. These have developed from early bullet-type connectors into high-quality waterproof systems now in use. A popular choice for many years was the spade terminal. This is still a standard choice for connection to relays for example, but is now losing ground to the smaller blade terminals. Circular multi-pin connectors are used in many cases; the pins varying in size from 1mm to 5mm. With any type of multi-pin connector an offset slot or similar is used to prevent incorrect connection.

Protection Protection against corrosion of the connector is provided in a number of ways. Earlier methods included applying suitable grease to the pins to repel water. It is now more usual to use rubber seals to protect the terminals although a small amount of contact lubricant can still be used. Many multi-connectors use some kind of latch to prevent not only individual pins working loose but also to ensure that the complete plug and socket is held securely.

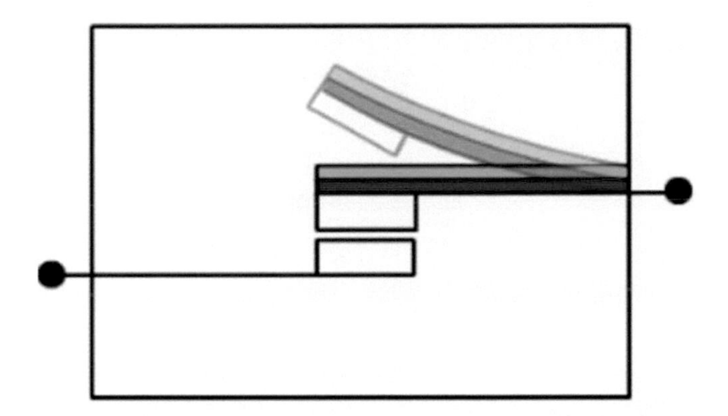

Figure 7.3 A bimetal strip is the main component

Figure 7.4 Terminals and connectors in use

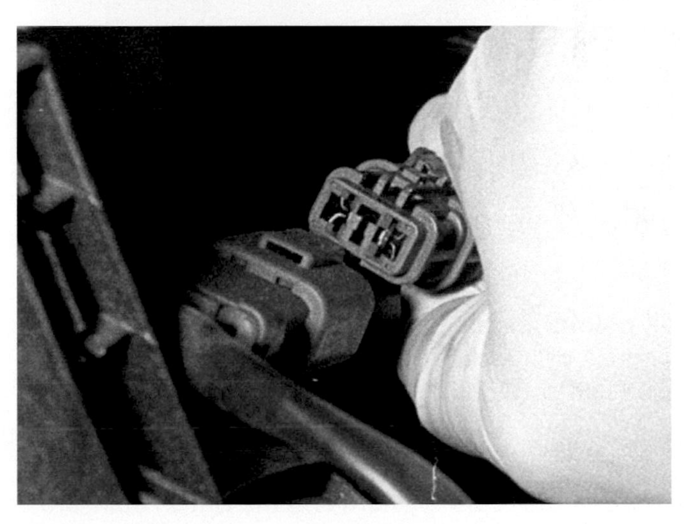

Figure 7.5 Waterproof connector block

Wires Cables or wires used for motor vehicle applications are usually copper strands insulated with PVC. Copper, beside its very low resistance, has ideal properties such as ductility and malleability. This makes it the natural choice for most electrical conductors. For the insulation, PVC is ideal. It not only has very high resistance, but also is very resistant to fuel, oil, water and other contaminants.

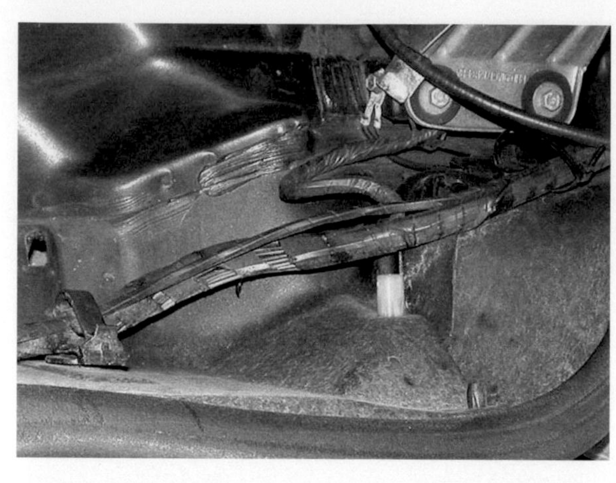

Figure 7.6 Cables in a wiring harness

Cable Size The choice of cable size depends on the current it will have to carry. The larger the cable used then the better it will be able to carry the current and supply all of the available voltage. However, it must not be too large or the wiring becomes cumbersome and heavy! In general, the voltage supply to a component must not be less than 90% of the system supply. Cable is available in stock sizes but a good 'rule of thumb' guide, is that one strand of 0.3mm diameter wire will carry 0.5 amps safely.

Figure 7.7 Heavy-duty cable

UK colour code The UK system uses twelve colours to determine the main purpose of the cable. Tracer colours further define its use. The main colours used and some other examples are given in the table.

A 'European' system used by Ford, VAG, BMW and other manufacturers is based broadly on the following table. Please note that there is no connection between the 'Euro' system and the British standard colour codes. In particular, note the use of the colour brown in each system!

A popular system is the terminal designation. This helps to ensure correct connections are made on the vehicle, particularly in after sales repairs. It is important however to note that the designations are not to identify individual wires but are to define the terminals of a device. Listed here are some of the popular numbers.

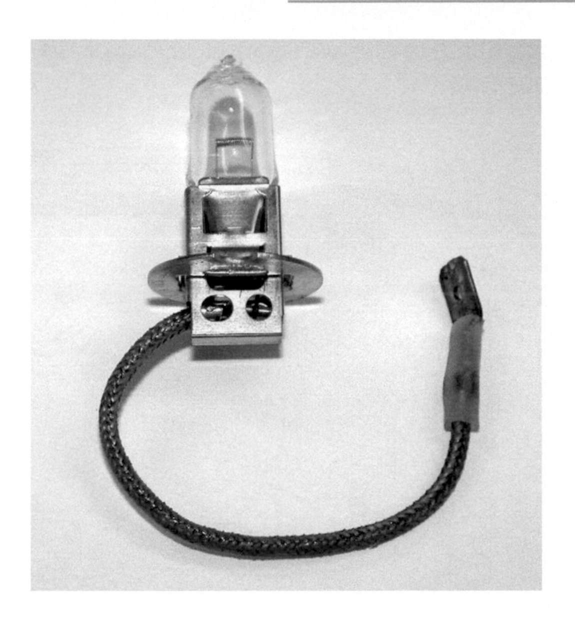

Figure 7.8 Light-duty cable

Symbols and circuit diagrams The selection of symbols shown here is intended as a guide to some of those in use. Many manufacturers use their own variation. The idea of a symbol is to represent a component in a very simple but easily recognisable form.

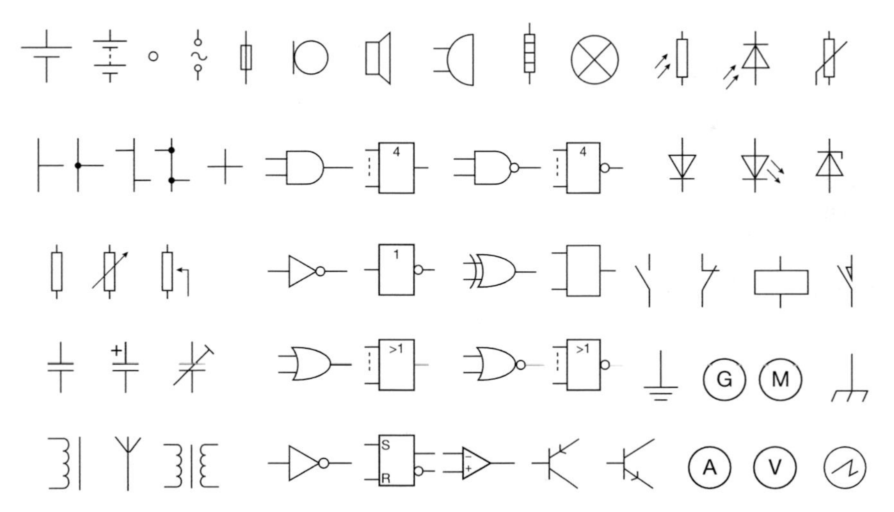

Figure 7.9 Symbols

The conventional type of diagram shows the electrical connections of a circuit but does not attempt to show the various parts in any particular order or position.

Layout circuit diagram A layout circuit diagram attempts to show the main electrical components in a position similar to those on the actual vehicle. Due to the complex circuits and the number of individual wires some manufacturers now use two diagrams, one to show electrical connections, and the other to show the actual layout of the wiring harness and components.

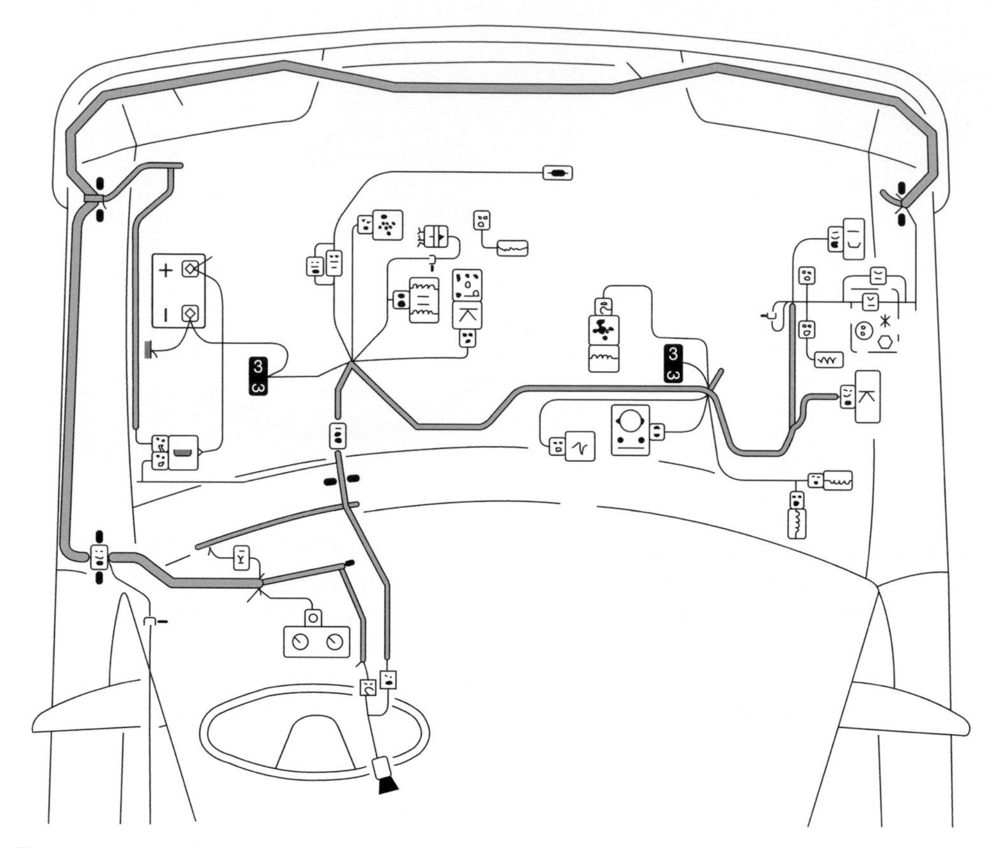

Figure 7.10 Layout circuit

Terminal circuit diagram A terminal diagram shows only the connections of the devices and not any of the wiring. The terminal of each device, which can be represented pictorially, is marked with a code. This code indicates the device terminal designation, the destination device code and its terminal designation and in some cases the wire colour code.

This diagram is laid out such as to show current flow from the top of the page to the bottom. These diagrams often have two supply lines at the top of the page marked 30 (main battery positive supply) and 15 (ignition controlled supply). At the bottom of the diagram is a line marked 31 (earth or chassis connection).

Shown here is a basic lighting circuit. Click on each switch in turn to make the circuit operate. Notice the effect of some switches being connected in series.

Describing electric circuit faults Three descriptive terms are useful when discussing electric circuits:

1 Open circuit – the circuit is broken and no current can flow.
2 Short circuit – a fault has caused a wire to touch another conductor and the current uses this as an easier way to complete the circuit.

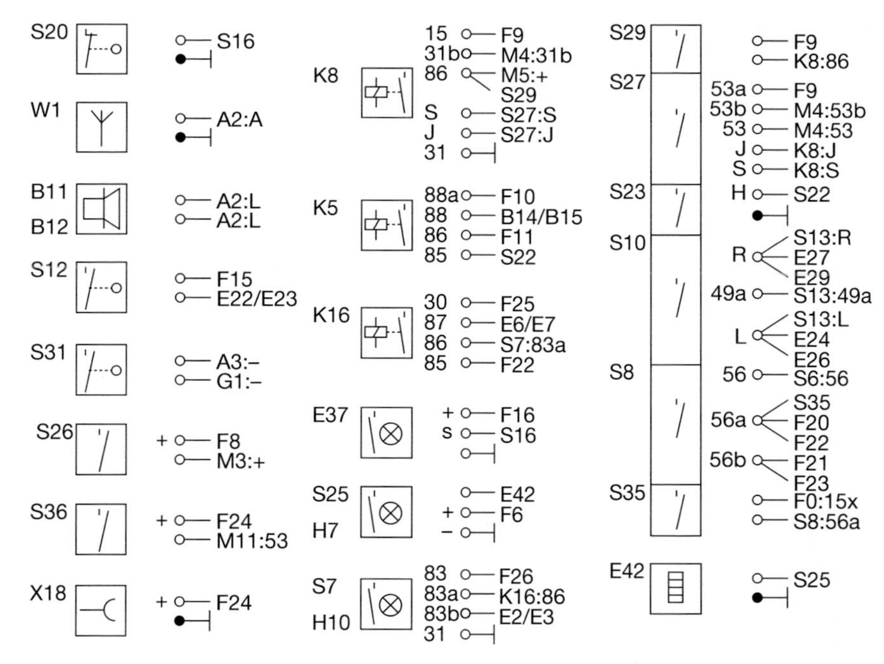

Figure 7.11 Terminal circuit

3 High resistance – a part of the circuit has developed a high resistance (such as a dirty connection), which will reduce the amount of current that can flow.

Figure 7.12 Short circuit

Figure 7.13 Open circuit

Look back over the previous section and write out a list of the key bullet points.

Make a simple sketch to show how one of the main components or systems in this section operates.

Remember! There is a multimedia version of this textbook that includes additional images and interactive features: www.automotivett.org

7.1.3 Lighting systems

Introduction Vehicle lighting systems are very important, particularly where road safety is concerned. If headlights were suddenly to fail at night and at high speed, the result could be serious. Remember, that lights are to see with, and to be seen by . . .

Lighting clusters Lights are arranged on a vehicle to meet legal requirements and to look good. Headlights, sidelights and indicators are often combined on the front. Taillights, stoplights, reverse lights and indicators are often combined at the rear.

Figure 7.14 Lights . . .

Figure 7.15 In different ways by . . .

Figure 7.16 Are positioned . . .

Figure 7.17 Different manufacturers

Bulbs The number, shape and size of bulbs used on vehicles is increasing all the time. A common selection is shown here. Most bulbs used for vehicle lighting are generally either conventional tungsten filament bulbs or tungsten halogen.

In the conventional bulb, the tungsten filament is heated to incandescence by an electric current. The temperature reaches about 23000C. Tungsten, or an alloy of tungsten, is ideal for use as filaments for electric light bulbs. The filament is normally wound into a 'spiralled spiral' to allow a suitable length of thin wire in a small space, and to provide some mechanical strength.

Tungsten halogen bulbs Almost all vehicles now use tungsten halogen bulbs for the headlights. The bulb will not blacken and therefore, has a long life. In normal gas bulbs, about 10% of the filament metal evaporates. This

Figure 7.18 Selection of bulbs

is deposited on the bulb wall. Design features of the tungsten halogen bulb prevent deposition. The gas in halogen bulbs is mostly iodine. The glass envelope is made from fused silicon or quartz.

Figure 7.19 Headlight bulb

Headlight reflectors The object of the headlight reflector is to direct the random light rays produced by the bulb into a beam of concentrated light, by applying the laws of reflection. Bulb filament position relative to the reflector is important, if the desired beam direction and shape are to be obtained.

Figure 7.20 Reflector

Reflector construction A reflector is a layer of silver, chrome or aluminium deposited on a smooth and polished surface such as brass or glass. Consider a mirror reflector that 'caves in' this is called a concave reflector. The centre point on the reflector is called the pole, and a line drawn perpendicular to the surface from the pole is known as the principal axis.

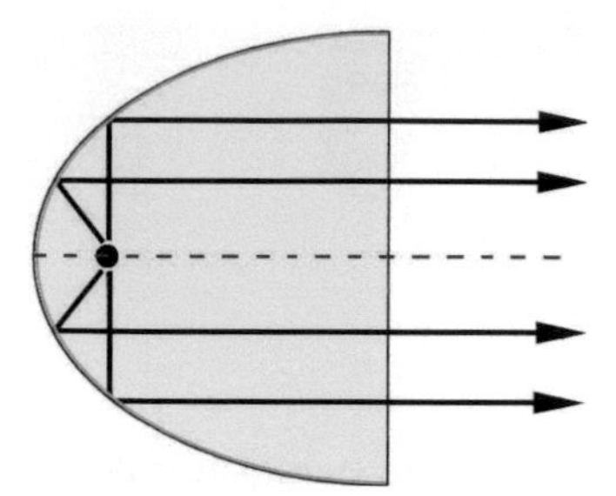

Figure 7.21 Concave reflector

Focused beam If a light source is moved along the principal axis, a point will be found where the radiating light produces a reflected beam parallel to the axis. This point is known as the focal point, and its distance from the pole is known as the focal length.

Light source at the focal point

Divergent and convergent beams If the filament is between the focal point and the reflector, the reflected beam will diverge – that is, spread outwards along the principal axis. If the filament is positioned in front of the focal point, the reflected beam will converge towards the principal axis.

The intensity of reflected light is strongest near the beam axis, except for the light cut off by the bulb itself. The intensity, therefore, dropping off towards the outer edges of the beam. A common type of reflector and bulb arrangement is shown here, where the dip filament is shielded. This gives a nice sharp cut – offline when on dip beam. It is used with asymmetric headlights.

Headlight lenses A good headlight should have a powerful far-reaching central beam, around which the light is distributed both horizontally and vertically in order to illuminate as great an area of the road surface as possible. The beam formation can be considerably improved by passing the reflected light rays through a transparent block of lenses. It is the function of the lenses

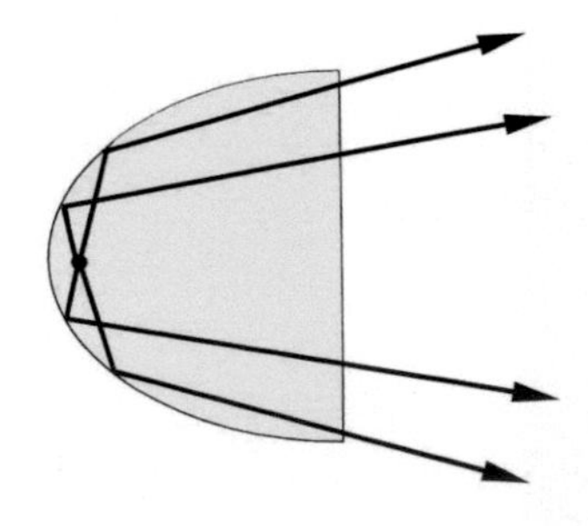

Figure 7.22 Light source behind the focal point

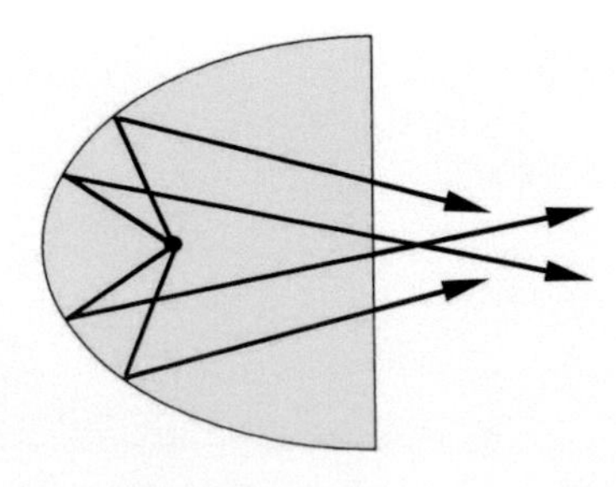

Figure 7.23 Light source in front of the focal point

to partially redistribute the reflected light beam and any stray light rays. This gives better overall road illumination.

Lenses Lenses work on the principle of refraction. The headlight front cover is≈the lens. It is divided up into a large number of small rectangular zones, each zone being formed optically in the shape of a concave flute or a combination of flute and prisms. Each individual lens element will redirect the light rays to obtain an improved overall light projection or beam pattern

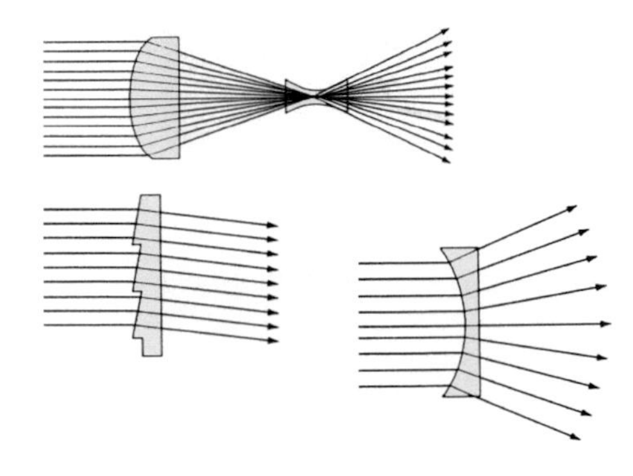

Figure 7.24 Headlight lens details

Complex shape reflectors Many headlights are now made with clear lenses, which means that all the direction of the light is achieved by the reflector. The clear lens does not restrict the light in any way. This makes the headlights more efficient as well as attractive.

Figure 7.25 Modern headlights

Other lights Sidelights, taillights, brake lights and others are relatively straightforward. Headlights present the most problems. This is because on dipped beam, they must provide adequate light for the driver, but not by dazzling other road users.

Headlight alignment The conflict between seeing and dazzling is very difficult to overcome. One of the latest developments, UV lighting, which is discussed later, shows some promise. The main requirement is that headlight alignment must be set correctly.

Figure 7.26 Rear lights

Figure 7.27 Beam setter in use

The function of a **levelling actuator** is to adjust the dipped or low beam in accordance with the load carried by the car. This will avoid dazzling oncoming traffic. Manual electric levelling actuators are connected up to a control on the dashboard. This allows the driver to adjust beam height.

Automatic static actuators adjust beam height to the optimum position in line with vehicle load conditions. The system includes two sensors (front and rear), which measure the attitude of the vehicle. An electronic module converts data from the sensors and drives two electric gear motors (or actuators) located at the rear of the headlamps, which are mechanically attached to the reflectors.

Create a word cloud for one or more of the most important screens or blocks of text in this section.

Light emitting diodes The main advantages of light emitting diodes (LEDs), when used for lighting, is that they have a typical rated life of over 50 000 hours. The environment in which vehicle lights have to survive is hostile.

Extreme variations in temperature and humidity, as well as serious shocks and vibration, have to be endured. LEDs are being developed in red, green and blue (RGB) groups. This will allow white light as well as other colours. The design possibilities for rear lights are therefore limitless.

Infrared lights Thermal imaging technology promises to make night driving visibly less hazardous. Infrared thermal-imaging systems are going to be fitted to cars. General Motors is now offering a system called 'Night Vision' as an option. After 'Night Vision' is switched on, 'hot' objects, including animals and people, show up as white in the thermal image. The image is projected onto the windscreen. On the vehicle, a camera unit sits in the centre of the car behind the front grille.

Figure 7.28 Night vision

High intensity discharge lamps (HIDs) are now being fitted to vehicles. They have the potential to provide illumination that is more effective, and new design possibilities for the front of a vehicle. The conflict between aerodynamic styling and suitable lighting positions is an economy/safety trade-off, which is undesirable.

The source of light in the gas discharge lamp is an electric arc. The actual discharge bulb used, is only about 10mm across. Two electrodes extend into the bulb, which is made from quartz glass. The gap between these electrodes is about 4mm. The bulb is filled with xenon gas. These lamps are sometimes described as high intensity discharge (HID).

If the HID system is used as a dip beam, the self-levelling lights are required because of the high intensities. Use, as a main beam may be a problem because of the on/off nature. An HID system for dip beam, which stays on all the time, is supplemented by a conventional main beam.

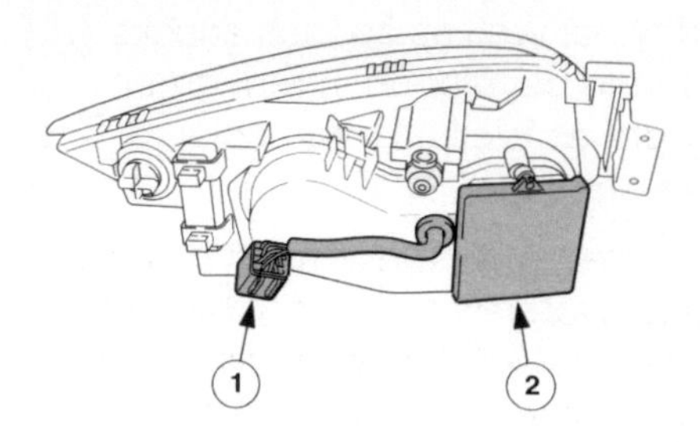

Figure 7.29 HID system headlamp

Summary Good lights are vital for safe driving at night. Interesting developments are taking place continuously. Many of the sophisticated systems, now being introduced on top of the range cars, will soon be available as standard options. Drive safely.

Look back over the previous section and write out a list of the key bullet points.

Now complete the multiple choice quiz associated with this topic/ subject area.

Use a library or the interactive web search tools to examine the subject in this section in more detail.

7.1.4 Stoplights and reverse lights

Introduction Stoplights, or brake lights, are used to warn drivers behind that you are slowing down or stopping. Reverse lights warn other drivers that you are reversing, or intend to reverse. The circuits are quite simple. One switch in each case operates two or three bulbs. A relay may be used.

The circuits for these two systems are similar. Shown here is a typical stoplight or reverse light circuit. Most incorporate a relay to switch on the lights, which is in turn operated by a spring-loaded switch on the brake pedal or gearbox. Links from the stoplight circuit to the cruise control system may be found. This is to cause the cruise control to switch off as the brakes are operated. A link may also be made to the antilock brake system.

Switches The circuits are operated by the appropriate switch. The reverse switch is part of the gearbox or gear change linkage. The stop switch is usually fitted so it acts on the brake pedal.

The diagram shown is the complete lighting circuit of a vehicle. The colour codes used are discussed in the basic electrical learning sections. However,

Figure 7.30 Stop and reverse lights form part of the rear light cluster

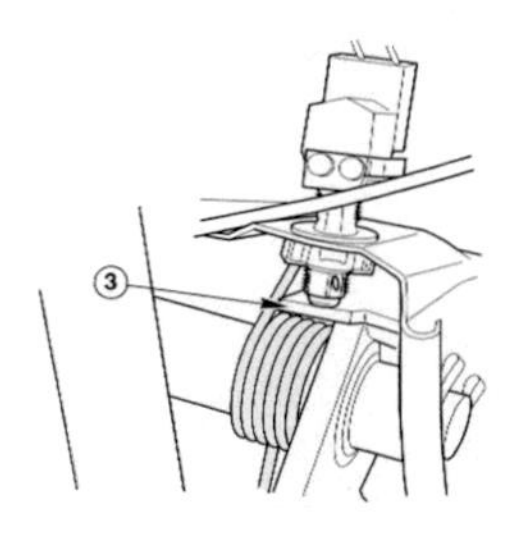

Figure 7.31 Stoplight switch

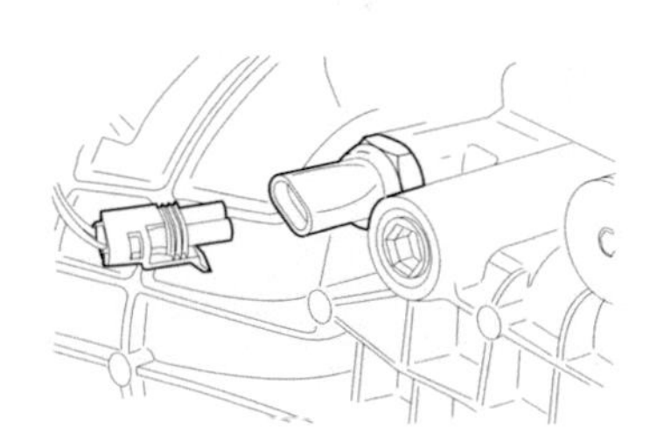

Figure 7.32 Reverse light switch

you can follow the circuit by looking for the labels on the wires. 'N' for example, means 'Brown' but this has no effect on how it works! Operation of part of this circuit is as described over the following screens:

The ignition must be on for these lights to operate. The reverse light switch gets its feed from fuse 16 on the GY wire. When the switch is operated, the supply is sent to the rear lamps on a GN wire. The switch is usually mounted on the gear change linkage or screwed into the gearbox.

The ignition must be on for these lights to operate. The brake or stoplight switch gets its feed from fuse 16 on the GY wire. When the switch is operated, the supply is sent to the rear lamps on a GP wire. A connection is also made to the centre high mounted stoplight. The switch is usually mounted on the pedal box above the brake pedal.

Light emitting diodes Light emitting diodes (LEDs) are more expensive than bulbs. However, the potential savings in design costs due to long life, sealed units being used and greater freedom of design, could outweigh the extra expense. LEDs are ideal for stoplights. A further advantage is that they illuminate quicker than ordinary bulbs. This time is approximately the difference between 130mS for the LEDs, and 200mS for bulbs. If this is related to a vehicle brake light at motorway speeds, then the increased reaction time equates to about a car length. This is potentially a major contribution to road safety.

Centre high mounted stop lamps An LED centre high mounted stop lamp (CHMSL) illuminates faster than conventional incandescent lamps, improving driver response time and providing extra braking distance. Due to their low height and reduced depth, LED CHMSLs can be easily harmonised with all vehicle designs. They can be mounted inside or integrated into the exterior body or spoiler.

Figure 7.33 CHMSL

Summary Reverse lights are operated by a simple on/off gearbox switch. Stoplights are operated by a simple on/off switch on the pedal box. Both circuits operate in much the same way. High mounted stoplights are now quite common, many of these using LEDs.

Make a simple sketch to show a brake light circuit using a relay.Keep a hard copy or save it electronically.

Look back over the previous section and write out a list of the key bullet points.

Construct a crossword puzzle using important words from this section. Hint: use the ATT glossary where you can copy the words and definitions (clues!). About 20 words is a good puzzle.

7.2 Electrical systems puzzles

7.2.1 Cryptograms

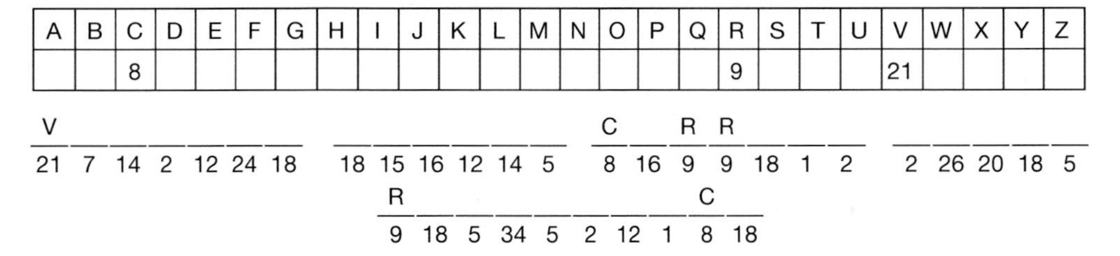

A	B	C	D	E	F	G	H	I	J	K	L	M	N	O	P	Q	R	S	T	U	V	W	X	Y	Z
		8															9				21				

V _ _ _ _ _ _   _ _ _ _ _ _   C _ R R _ _ _   _ _ _ _ _

21 7 14 2 12 24 18   18 15 16 12 14 5   8 16 9 9 18 1 2   2 26 20 18 5

R _ _ _ _ _ _ _ C _

9 18 5 34 5 2 12 1 8 18

Cryptogram 7.1 Charge the battery:

7.2.2 Anagrams

Rice City Let	Trice Icy Let
Lyric Tic Tee	Tic Cry Elite

Anagrams 7–1 Hurts to the touch: what word did I use to get these anagrams?

7.2.3 Word search

V	F	R	W	U	A	G	S	A	Z	D	J	P	E	R	
Q	E	H	X	R	D	T	K	U	O	F	M	B	L	O	
T	O	Z	Y	L	I	K	A	Q	O	W	I	E	E	T	
O	R	J	Z	U	P	D	R	D	F	O	L	Z	C	C	
Q	R	A	C	L	O	N	N	R	S	E	N	B	T	U	
Q	O	R	N	K	R	L	O	J	C	F	U	V	R	D	
W	I	U	R	S	K	T	S	T	P	T	H	O	I	N	CABLE
C	B	S	V	V	F	X	R	N	O	L	P	B	C	O	CIRCUITS
E	L	E	C	T	R	O	M	A	G	N	E	T	A	C	CONDUCTOR
O	X	A	N	Z	N	J	R	C	I	C	N	A	L	J	ELECTRICAL
A	K	A	U	I	Q	C	H	M	Z	K	A	Y	R	P	ELECTROMAGNET
M	G	N	C	Y	Z	K	I	W	E	K	T	B	R	Y	ELECTRONIC
E	C	N	A	T	S	I	S	E	R	R	E	V	L	L	ENERGY
T	M	Y	I	R	H	Y	K	S	U	C	F	N	V	E	RESISTANCE
Q	X	A	X	X	R	W	C	E	N	E	R	G	Y	N	TRANSFORMER

Word search 7.1 Electrical and electronic principles

7.2.4 Crossword

Across

4 A material that can be made to conduct or block electricity; a diode for example

8 The gas in an air conditioning system

9 A powerful electric machine to rotate the engine so it then runs on its own (7,5)

10 Electromotive force or potential difference

Down

1 A description of connecting positive and negative connected correctly

2 Starting a vehicle engine with the aid of a slave battery or charger (4,5)

3 The relationship between volts, current and resistance (5,3)

5 The opposition to electrical current flow in a conductor

Remember! There is a multimedia version of this textbook that includes additional images and interactive features: www.automotivett.org

Across

11 Electrical connection points

12 An electrical generator attached to and driven by a motor vehicle engine

14 A tachometer that indicated revolutions per second

15 A chemical device that will give out an electrical current when connected into a circuit

17 Electronic Control Unit (1,1,1)

23 Applied to an electrical circuit that is broken or not connected (4,7)

26 A natural force found in iron and some other materials

27 Electromagnetic switch used in electric circuits to reduce load on smaller switches

29 Movement of electricity

31 Equipment used to check the alignment of vehicle headlights (4,6)

32 Electrical unit of resistance

33 This fault allows a current to travel along a different path from the one intended (5,7)

34 Machine for converting mechanical energy into electrical energy

35 The power rating of an electrical device

Down

6 A safety device in an electrical circuit

7 Electrical circuit with all components connected in their own individual circuits (8,7)

9 Device that converts audio signals to sounds that humans can hear

13 Restricts the flow of electricity or heat

16 Formed by an electrical winding

18 A system of numbering using 0 and 1 and used in computer operating systems

19 A switch used to select main beam or low beam headlights (3,6)

20 The collection of vehicle cables wound together with an insulation tape

21 Any electric circuit having all elements joined in a sequence (6,7)

22 The tungsten element in a bulb that glows to produce light

24 Two plates separated by an insulator capable of holding a charge

25 A liquid that conducts electricity

28 A process where digital signals share a single path

30 Electrical unit for measuring the flow of electricity in a circuit

Crossword 7.1 Electrical and electronic principles and electrical systems

CHAPTER 8

Chassis systems

8.1 Principles of light vehicle steering and suspension systems

Together with the multimedia resources, this section is ideal material for students working towards the IMI Awards unit EL05, the City and Guilds 3902–008 unit and all other similar foundation or introduction level qualifications.

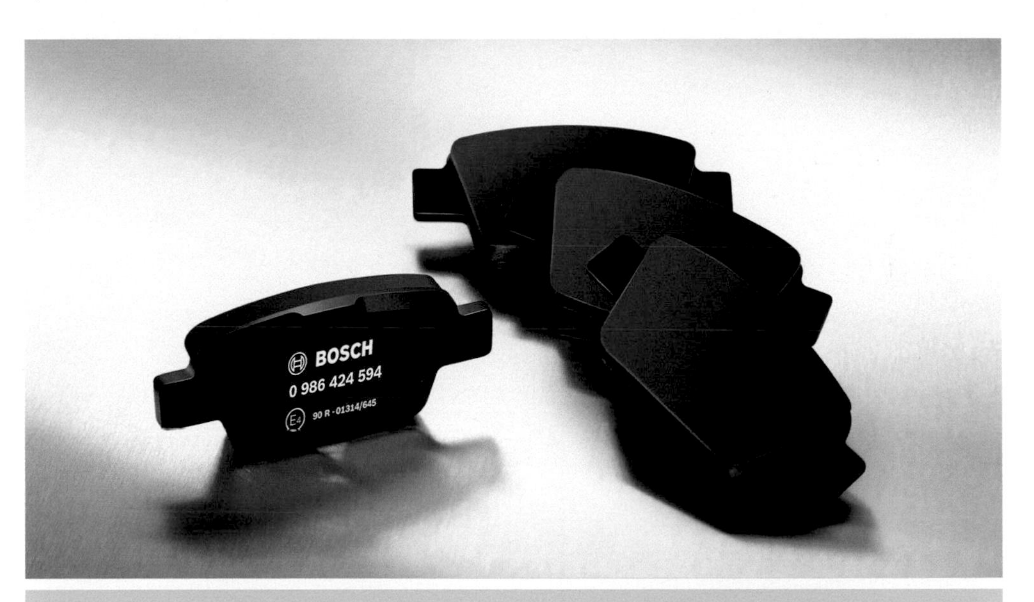

After successful completion of this section you will be able to show you have achieved these objectives:

- Be able to work safely.
- Know about steering systems.
- Be able to carry out simple checks to steering systems.
- Know about suspension systems.
- Be able to carry out simple checks on suspension systems.

Automotive Technician Training: Entry Level 3. 978-0-415-72040-3.
© Tom Denton. Published by Taylor & Francis. All rights reserved.

8.1.1 What you must know about steering and suspension:

Functions of the suspension system are to: provide a safe and pleasant ride for the car occupants, provide positive steering and handling of the vehicle, and enable the driver to be in full control of the vehicle under all conditions

Functions of steering systems are to: enable the driver to control the path of the vehicle, and be light and easy to operate. The driver converts effort into force to turn the wheels: rotary movement at the steering wheel turned into linear movement at the wheels, gearing is used to decrease drivers' effort

Always wear suitable PPE and use tools and equipment safely and correctly when doing practical work on any vehicle

The main components of light vehicle suspension system include: telescopic dampers, leaf springs, coil springs, torsion bars, McPherson strut, anti-roll bars and suspension arms

Check a steering rack bellows for damage by visual inspection for damage, leaks and fouling

Check and top-up power assisted steering fluid level using appropriate means, selecting correct fluid by referring to data and clearing up any spillage

Check the front-wheel alignment using simple equipment to include: calibration, appropriate pre-checks on vehicle and positioning, correct location of equipment on vehicle and correct readings obtained

The main components of a light vehicle steering system include: steering wheel, steering column, steering gear (rack) and track rods

8.1.2 Suspension

Introduction The suspension system is the link between the vehicle body and the wheels. Its purpose is to:

- Locate the wheels while allowing them to move up and down, and steer.
- Maintain the wheels in contact with the road and minimise road noise.
- Distribute the weight of the vehicle to the wheels.
- Reduce vehicle weight as much as possible – in particular the unsprung mass
- Resist the effects of steering, braking and acceleration.
- Work in conjunction with the tyres and seat springs to give acceptable ride comfort.

Figure 8.1 Suspension plays a key role

Compromise The previous list is difficult to achieve completely, so some sort of compromise has to be reached. Because of this, many different methods have been tried, and many are still in use. Keep these requirements in mind, and it will help you to understand why some systems are constructed in different ways.

Further in suspension A vehicle needs a suspension system to cushion and damp out road shocks. This provides comfort to the passengers and prevents damage to the load and vehicle components. A spring between the wheel and the vehicle body allows the wheel to follow the road surface. The tyre plays an important role in absorbing small road shocks. It is often described as the primary form of suspension. The vehicle body is supported by springs located between the body and the wheel axles. Together with the damper, these components are referred to as the suspension system.

Figure 8.2 Suspension system

Effect of suspension As a wheel hits a bump in the road, it is moved upwards with quite some force. An unsprung wheel is affected only by gravity, which will try to return the wheel to the road surface. However, most of the energy will be transferred to the body. When a spring is used between the wheel and the vehicle body, most of the energy in the bouncing wheel is stored in the spring and not passed to the vehicle body. The vehicle body will only moves upwards through a very small distance compared to the movement of the wheel.

Springs These parts of the suspension system take up the movement or shock from the road. The energy of the movement is stored in the spring. The actual spring itself can be in many different forms, ranging from a steel coil to a pressurised chamber of nitrogen. Soft springs provide the best comfort, but stiff springs can be better for high performance. Vehicle springs and suspension therefore are made to provide a compromise between good handling and comfort.

Make a simple sketch to show different types of springs. Keep a hard copy or save it electronically.

Dampers or shock absorbers The energy stored in the spring after a bump, has to be got rid of or else the spring would oscillate (bounce up and down).

Figure 8.3 Coil spring

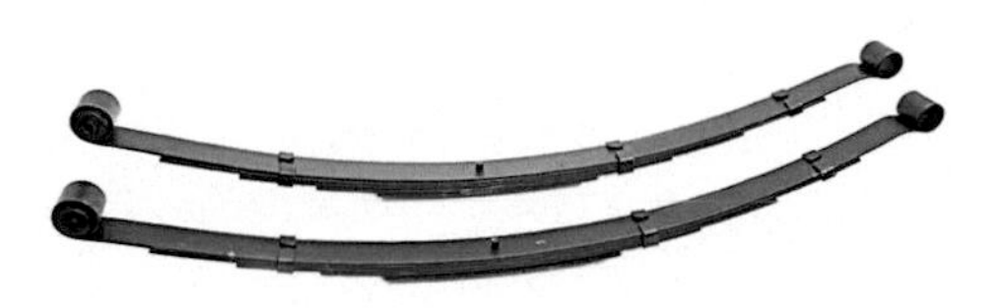

Figure 8.4 Leaf spring

Figure 8.5 Gas spring

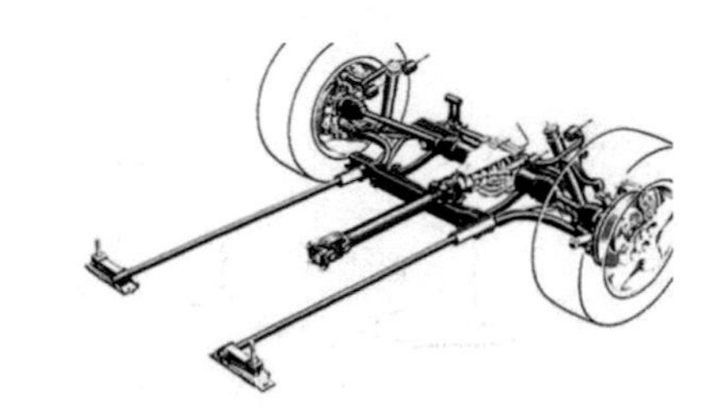

Figure 8.6 Torsion bar spring

The damper damps down these oscillations by converting the energy from the spring into heat. If working correctly the spring should stop moving after just one bounce and rebound. Shock absorber is a term, which is often used, to describe a damper.

Strut The combination of a coil spring with a damper inside it, between the wheel stub axle and the inner wing, is often referred to as a strut. This is a very popular type of suspension.

Wishbone A wishbone is a triangular shaped component with two corners hinged in a straight line on the vehicle body. The third corner is hinged to the moving part of the suspension.

Figure 8.7 Telescopic damper

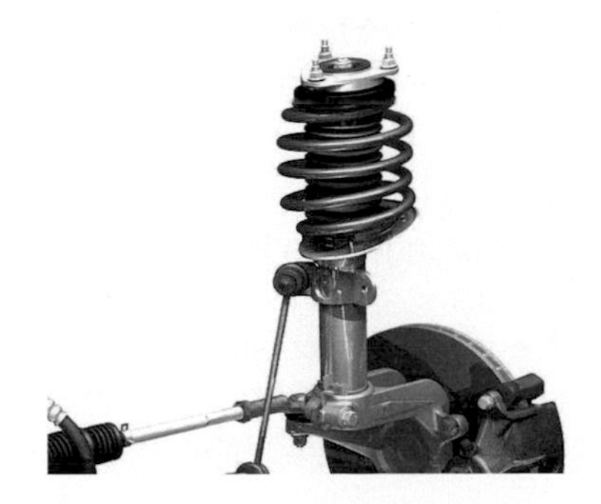

Figure 8.8 McPherson strut

Figure 8.9 Front suspension wishbone

Figure 8.10 Rubber stop

Figure 8.11 Heavy vehicle axle

Bump stop When a vehicle hits a particularly large bump, or if it is carrying a heavy load, the suspension system may bottom out (reach the end of its travel). The bump stop, usually made of rubber, prevents metal-to-metal contact, which would cause damage.

Link A link is a very general term, which is used to describe a bar or other similar component that holds or controls the position of another component. Other terms may be used such as tie-bar or tie-rod.

Beam axle This is a solid axle from one wheel to the other. It is not now used on the majority of light vehicles. However, as it makes a very strong construction, it is still common on heavy vehicles.

Gas/fluid suspension The most common types of spring are made from steel. However, some vehicles use pressurised gas as the spring (think of a balloon or a football). On some vehicles, a connection between wheels is made using fluid running through pipes from one suspension unit to another.

Independent suspension

Independent front and rear suspension (IFS/IRS) was developed to meet the demand for improved ride quality and handling. The main advantages of independent suspension are as follows:

- When one wheel is lifted or drops, it does not affect the opposite wheel.
- The unsprung mass is lower; therefore, the road wheel stays in better contact with the road.
- Problems with changing steering geometry are reduced.
- More space for the engine at the front.
- Softer springing with larger wheel movement is possible.

Figure 8.12 Gas suspension unit

Anti-roll bar The main purpose of an anti-roll bar is to reduce body roll on corners. The anti-roll bar can be thought of as a torsion bar. The centre is pivoted on the body and each end bends to make connection with the suspension/wheel assembly. When the suspension is compressed on both sides, the anti-roll bar has no effect because it pivots on its mountings. As the suspension is compressed on just one side, a twisting force is exerted on the anti-roll bar. Part of this load is transmitted to the opposite wheel, pulling it upwards. This reduces the amount of body roll on corners.

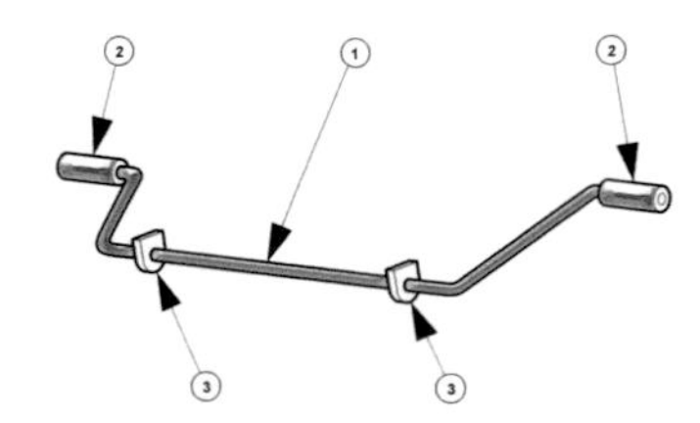

Figure 8.13 Shape of an anti-roll bar

Panhard rod The Panhard rod was named after a French engineer. Its purpose is to link a rear axle to the body. The rod is pivoted at each end to allow movement. It takes up lateral forces between the axle and body thus removing load from the radius arms. The radius arms have now to only transmit longitudinal forces.

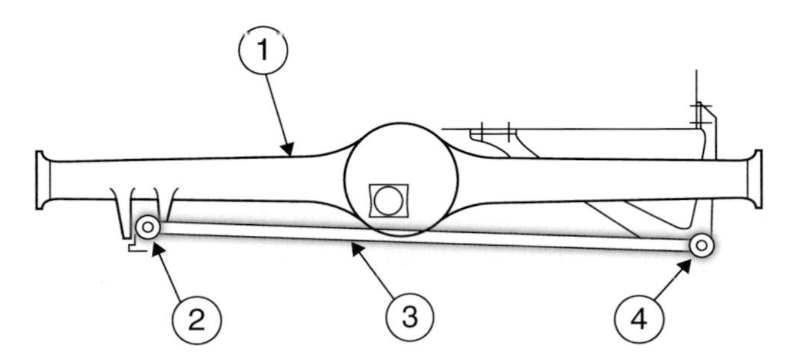

Figure 8.14 Rear axle with Panhard rod

Summary A wide variety of suspension systems and components are used. Engineers strive to achieve optimum comfort and handling. However, these two main requirements are often at odds with each other. As is common with all vehicle systems, electronic control is one-way developments are now being made.

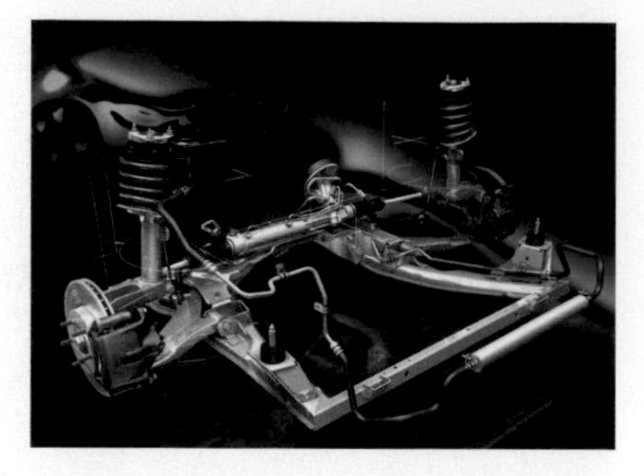

Figure 8.15 Suspension continues to develop

Look back over the previous section and write out a list of the key bullet points.

Create a mind map to illustrate the important features of a component or system in this section.

Introduction The requirements of the springs can be summarised as follows:

- Absorb road shocks from uneven surfaces.
- Control ground clearance and ride height.
- Ensure good tyre adhesion.
- Support the weight of the vehicle.
- Transmit gravity forces to the wheels.

There are a number of different types of spring in use on modern vehicles.

Coil springs Although modern vehicles use a number of different types of spring medium, the most popular is the coil (or helical) spring. Coil or helical springs used in vehicle suspension systems, are made from round spring steel bars. The heated bar is wound on a special former and then heat-treated, to obtain the correct elasticity (springiness). The spring can withstand any compression load but not side thrust. It is also difficult for a coil spring to resist braking or driving thrust. Suspension arms are used to resist these loads.

Independent suspension systems Coil springs are generally used with≈independent suspension systems; the springs are usually fitted on each side of the vehicle, between the stub axle assembly and the body. The spring remains in the correct position because recesses are made in both the stub axle assembly and body. The spring is always under compression due to the weight of the vehicle and hence holds itself in place.

Figure 8.16 Coil spring in position

Figure 8.17 Coil spring upper fitting

Figure 8.18 Coil spring lower fitting

Coil spring features The coil spring is a torsion bar wound into a spiral. It can be progressive if the diameter of the spring is tapered conically. A coil spring cannot transmit lateral or longitudinal forces, hence the need for links or arms. It produces little internal damping. No maintenance is required and high travel is possible.

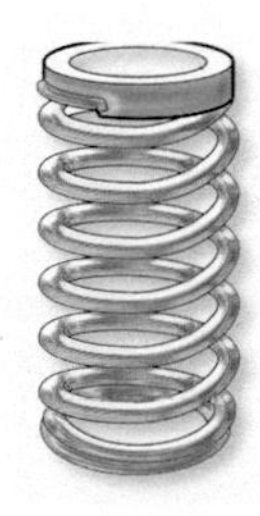

Figure 8.19 Details of a coil spring

Leaf springs The leaf spring can provide all the control for the wheels during acceleration, braking, cornering, and general movement caused by the road surface. They are used with fixed axles. Leaf springs can be described as:

- Laminated or multi-leaf springs
- Single leaf or mono-leaf springs.

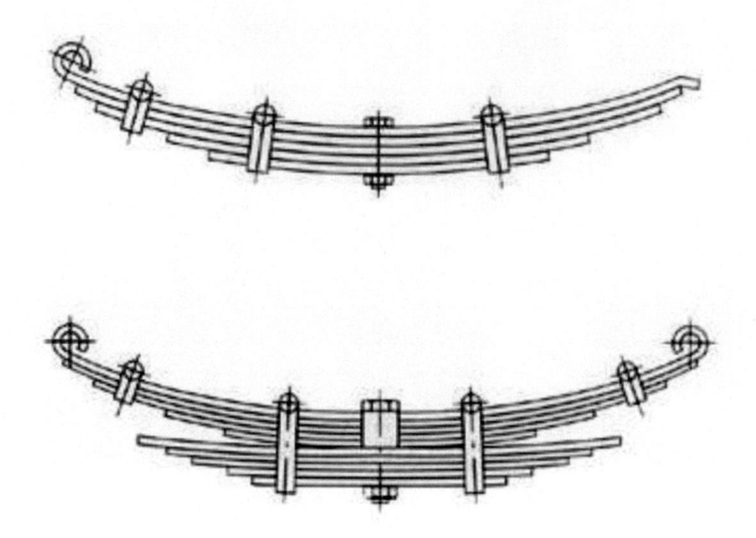

Figure 8.20 Laminated springs

Multi-leaf spring The multi-leaf spring was widely used at the rear of cars and light vehicles, and is still used in commercial vehicle suspension systems. It consists of a number of steel strips or leaves placed on top of each other and then clamped together. The length, cross-section and number of leaves are determined by the loads carried.

Figure 8.21 Commercial vehicle leaf spring

Leaf spring fixings The top leaf is called the main leaf and each end of this leaf is rolled to form an eye. This is for attachment to the vehicle chassis or body. The leaves of the spring are clamped together by a bolt or pin known as the centre bolt. The spring eye allows movement about a shackle and pin at the rear, allowing the spring to flex. The vehicle is pushed along by the rear axle through the front section of the spring, which is anchored, firmly to the fixed shackle on the vehicle chassis or body. The curve of leaf springs straightens out when a load is applied to it, and its length changes.

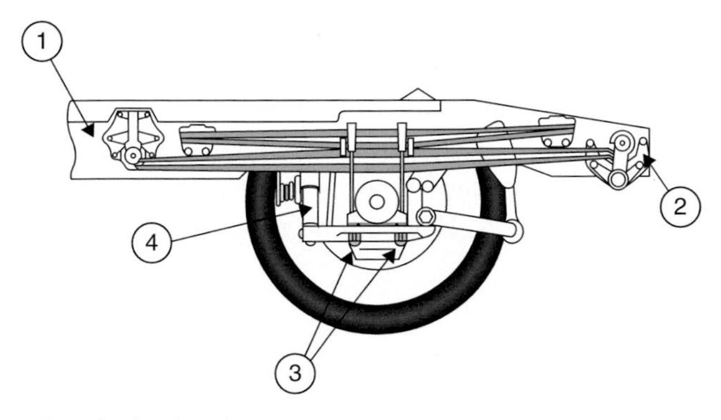

Figure 8.22 Details of a leaf spring

Shackles Because of the change in length as the spring moves, the rear end of a leaf spring is fixed by a shackle bolt to a swinging shackle. As the road wheel passes over a bump, the spring is compressed and the leaves slide over each other. As it returns to its original shape, the spring forces the wheel back in contact with the road. The leaf spring is usually secured to the axle by means of U bolts. As the leaves of the spring move, they rub together. This produces interleaf friction, which has a damping effect.

Single leaf spring A single leaf spring, as the name implies, consists of one uniformly stressed leaf. The spring varies in thickness from a maximum at the centre to a minimum at the spring eyes. This type of leaf spring is made to work in the same way as a multi-leaf spring. Advantages of this type of spring are:

- simplified construction
- constant performance over a period, because interleaf friction is eliminated
- reduction in unsprung mass.

Torsion bars This type of suspension uses a metal bar, which provides the springing effect as it is twisted. It has the advantage that the components do not take up too much room. The torsion bar

Figure 8.23 Tapered single leaf

Remember! There is a multimedia version of this textbook that includes additional images and interactive features: www.automotivett.org

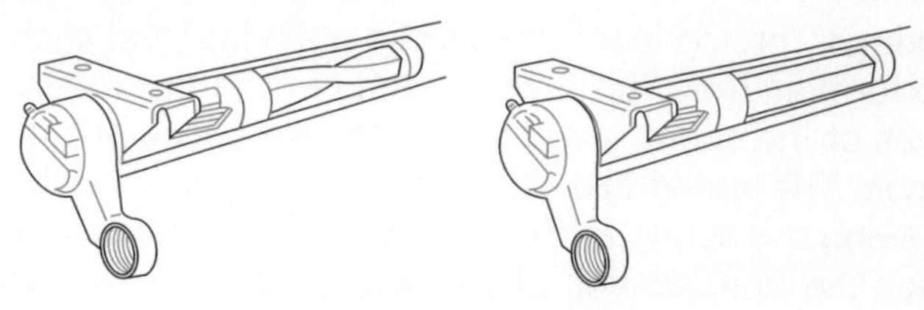

Figure 8.24 Torsion bar in a guide tube

can be round or square section, solid or hollow. The surface must be finished accurately to eliminate pressure points, which may cause cracking and fatigue failure. They can be fitted longitudinally or laterally

Torsion bar features Torsion bars are maintenance free but can be adjusted. They transmit longitudinal and lateral forces and have low mass. However, they have limited self-damping. Their spring rate is linear rate and life may be limited due to fatigue.

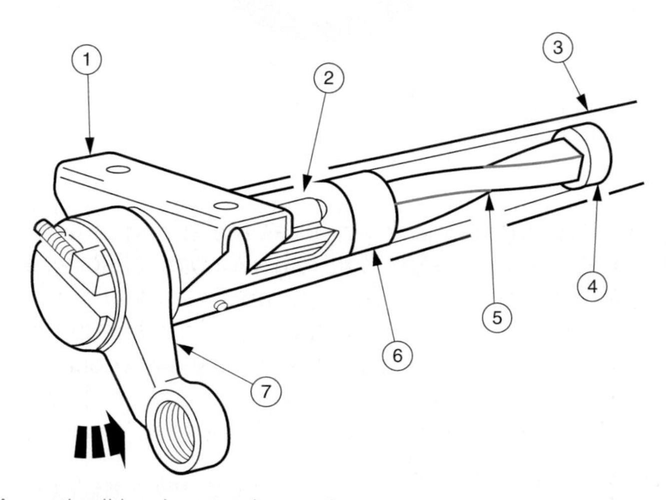

Figure 8.25 An anti-roll bar is a torsion spring

Summary Suspension springs can be made from a variety of materials and in many different ways. The most common is the coil spring. This is because it has many advantages and is reasonably inexpensive.

Look back over the previous section and write out a list of the key bullet points.

Use the interactive media search tools to look for pictures and videos to examine the subject in this section in more detail.

Dampers introduction As a spring is deflected, energy is stored in it. If the spring is free to move, the energy is released in the form of oscillations, for a short time, before it comes to rest. This principle can be demonstrated by flicking the end of a ruler placed on the edge of a desk. The function of the damper is to absorb the stored energy, which reduces the rebound oscillation. A spring without a damper would build up dangerous and uncomfortable bouncing of the vehicle.

Hydraulic dampers Hydraulic dampers are the most common type used on modern vehicles. They work on the principle of forcing fluid through small holes. In a hydraulic damper, the energy in the spring is converted into heat. This is caused as the fluid (a type of oil) is forced rapidly through small holes (orifices). The oil temperature in a damper can reach over 150OC during normal operation. As an example think of using a hand oil pump and how hard it is to make the oil flow quickly.

Damper functions The functions of a damper can be summarised as follows:

- Ensure directional stability.
- Ensure good contact between the tyres and the road.
- Prevent build-up of vertical movements.
- Reduce oscillations.
- Reduce wear on tyres and chassis components.

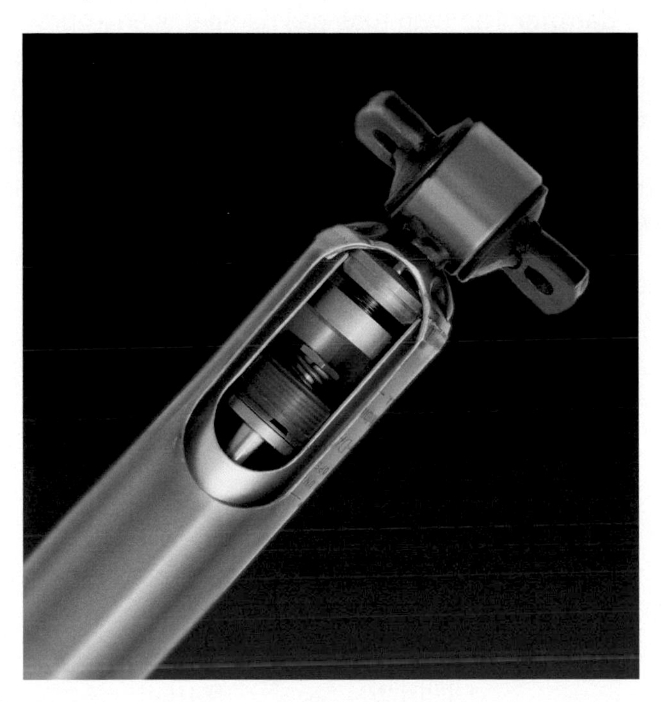

Figure 8.26 Damper

Single tube telescopic damper This is often referred to as a gas damper. However, the damping action is still achieved by forcing oil through a restriction. The gas space behind a separator piston is to compensate for the changes in cylinder volume, which is caused as the piston rod moves. The gas is at a pressure of about 25 bar.

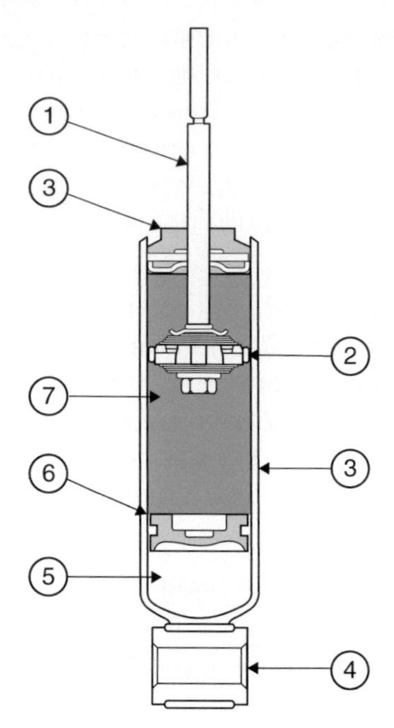

Figure 8.27 Single tube system

Twin tube gas damper The twin tube gas damper is an improvement on the well-used twin-tube system. The gas cushion is used in this case to prevent oil foaming. The gas pressure on the oil prevents foaming, which in turn ensures constant operation under all operating conditions. Gas pressure is set at about 5 bar. If bypass grooves are machined in the upper half of the working chamber, the damping rate can be made variable. With light loads the damper works in this area with a soft damping effect. When the load is increased the piston moves lower down the working chamber away from the grooves resulting in full damping effect.

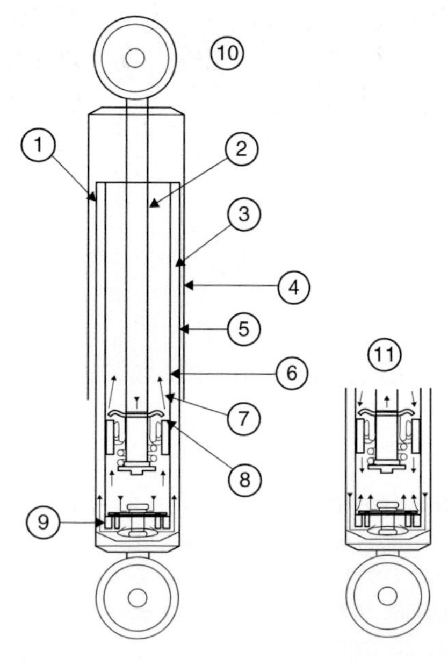

Figure 8.28 Twin tube gas system

Summary Dampers (or shock absorbers) are used to prevent the suspension springs oscillating. This improves handling, comfort and safety.

Figure 8.29 Suspension in action

Look back over the previous section and write out a list of the key bullet points.

Make a simple sketch to show how one of the main components or systems in this section operates.

8.1.3 Check damper operation

Damper operation Dampers (or shock absorbers) are an important part of the suspension system. They are quite easy to test. As you press on one corner of the vehicle, it should bounce back just past the start point and then return to the rest position. Repeat on all four corners. Make sure you press down on a strong section of the vehicle – some body panels are easily distorted!

Suspension bush condition All the rubber bushes used for mounting suspension components can be tested with a simple lever. This will show up excessive movement, cracks or separation of rubber bushes.

Trim height Trim height, is a measurement usually taken from the wheel centre, to a point on the car above. It is available in most data books.

Summary System performance checks are often quite simple. However, they are important. Cars are used at high speed and sudden breakdowns can be dangerous. The systems should therefore function correctly at all times.

Look back over the previous section and write out a list of the key bullet points here:

8.1.4 Steering

Development of steering systems The development of steering systems began before cars were invented. On early cars, the entire front axle was steered by way of a pivot (fifth wheel) situated in the centre of the vehicle. The steering accuracy was not very good, there was a serious risk of overturning and the tyre wear was significant.

Ackermann In 1817, Rudolf Ackermann patented the first stub axle steering system in which each front wheel was fixed to the front axle by a joint. This

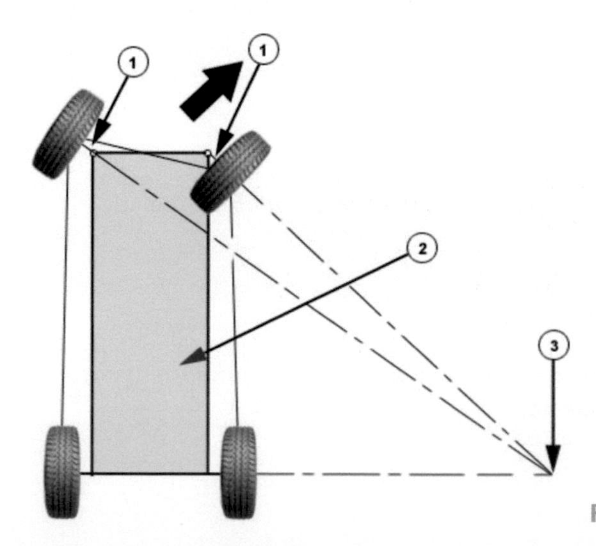

Figure 8.30 Ackermann steering

made it possible to cover a larger curve radius with the wheel on the outside of the curve than with the front wheel on the inside of the curve.

Rack and pinion steering Rack and pinion steering was developed at an early age in the history of the car. However, this became more popular, when front-wheel drive was used more, since it requires little space and production costs are lower. The first hydraulic power steering was produced in 1928. However, since there was no great demand for this until the fifties the development of power steering systems stagnated.

Figure 8.31 Steering rack

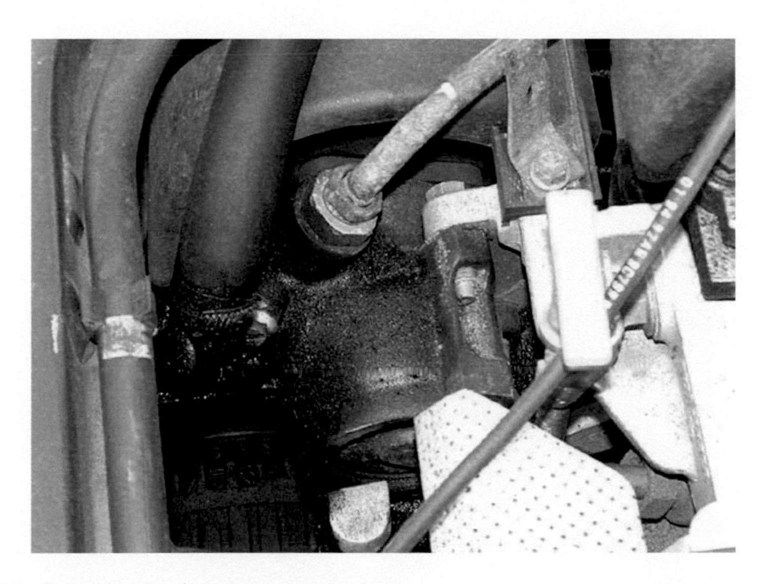

Figure 8.32 Power steering pump

Power steering systems Increasing standards of comfort stimulated the demand for power steering systems. Speed-sensitive or variable-assistance power steering (VAPS) systems were developed using electronic controls. These represent the latest major innovation to the steering system in production vehicles. The demand for safety and comfort will lead to further improvements in steering systems.

The necessity for steering systems Motor vehicles are generally steered via the front wheels, the rear wheels following the front wheels on a smaller radius. With motor vehicles, two factors have to be taken into account:

- dead weight or axle loading; and
- steered wheels' contact area.

In order to overcome the friction forces more easily, many different types of steering gear have been developed. Power steering in particular, reduces the effort required and increases the safety and comfort. Steering systems must be capable of:

- Automatically returning the steered front wheels to the straight-ahead position after cornering (self-centring action).
- Translating the steering wheel rotation so that only about two rotations of the steering wheel are necessary for a steering angle of about 40°.

Steering and suspension Steering and suspension must always be regarded as a unit. If the suspension system is not working correctly, it will have a considerable influence on the vehicle's steering characteristics. For example, defective shock absorbers or dampers reduce the wheel contact with the

Figure 8.33 Steering components

road, limiting the ability to steer the vehicle. The driving safety of a motor vehicle depends largely on the steering. Reliable steering at high speeds is required, together with easy manoeuvrability.

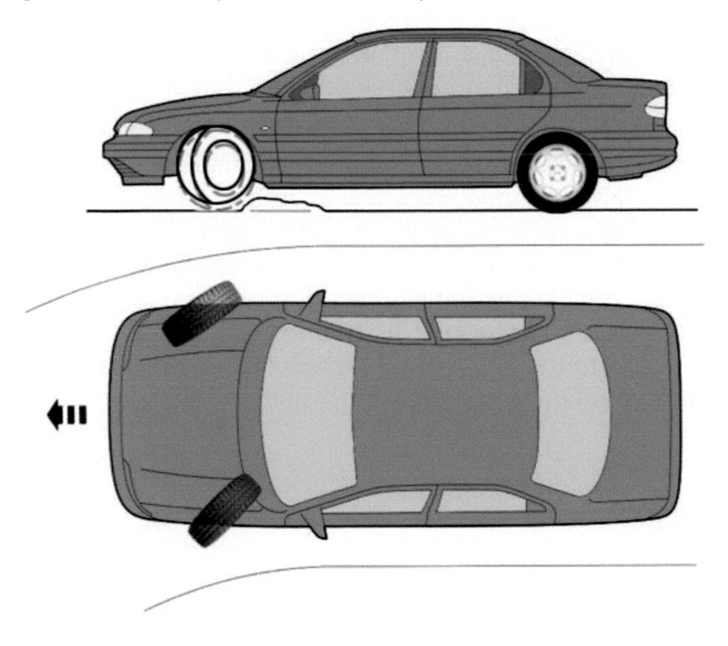

Figure 8.34 Suspension and steering interact

Manoeuvrability Crucial to the manoeuvrability of a motor vehicle is the turning circle, which in turn is directly dependent on the track circle. Designers strive for the smallest possible track and turning circle. The wheel housing should enclose the wheels as tightly as possible; however, sufficient clearance must be left so that the tyres do not rub when the wheels are turned.

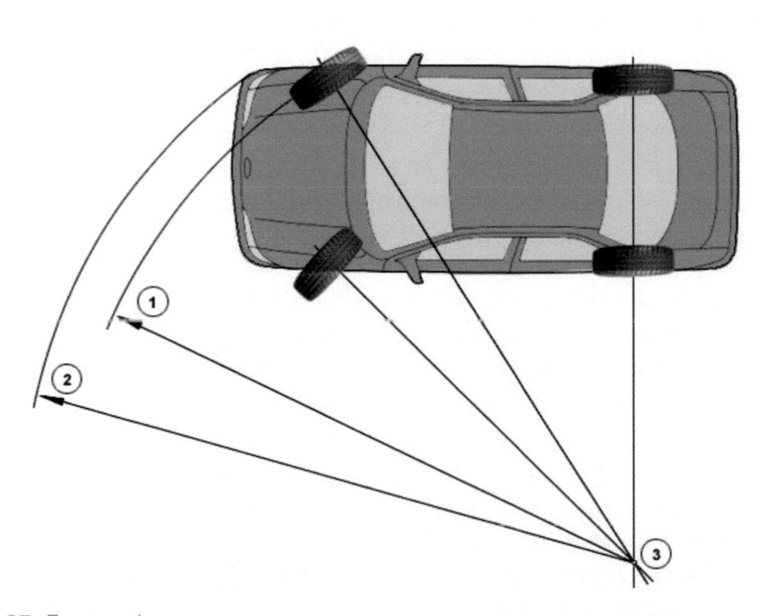

Figure 8.35 Front axle geometry

In this type of steering, the stub axle of the steered front wheel is swivelled about the steering axis. When steering, the wheelbase remains constant. The space between the steered wheels can be used for the installation of deep-seated components such as the engine. The low centre of gravity contributes to road handling characteristics. Even at large steering angles, the stability of the vehicle is maintained since the area of support is only slightly reduced.

The 'Trapezium' name is derived from the geometrical shape, which the two steering arms and the track rod form with the front axle. The stub axle and steering arm are firmly connected to one another. The stub axles are swivel mounted on the kingpins or in ball joints. Track rod and steering arms are movably connected to one another. When in the straight-ahead position, track rod and front axle are parallel. When cornering the stub axles are swivelled, thereby turning the front wheels. With the front wheels turned, the track rod is no longer parallel to the front axle. This results in the inside front wheel being turned more than the outside front wheel.

Look back over the previous section and write out a list of the key bullet points.

Create a mind map to illustrate the important features of a component or system in this section.

Construction of the steering system In order to transmit the steering movements of the driver to the wheels, several components are required. The steering movement is transmitted by way of the steering wheel, shaft, gear and linkage to the front wheels. The rotational movement of the steering wheel is transmitted via the steering shaft to the steering pinion in the steering gear. The steering shaft is supported in the steering column tube, which is fixed to the vehicle body.

Steering gear The steering gear translates (reduces) the steering force applied by the driver. It also converts the rotational movement of the steering wheel into push or pull movements of the track rods. The converted movement is transmitted to the linkage, which in turn moves the wheels in the desired steering direction. Track rods are required to transmit the steering movement from the steering gear to the front wheels. Different track rods are used depending on the type of front axle.

Ball joint Ball joints allow parts of the steering linkage to rotate about the longitudinal axis of the ball joint. They also allow limited swivel movements transversely to the longitudinal axis. The lubricated ball pivot is supported in steel cups or between preloaded plastic cups. A gaiter prevents lubricant losses. Ball joints are generally maintenance-free and must always be renewed if the gaiter is damaged.

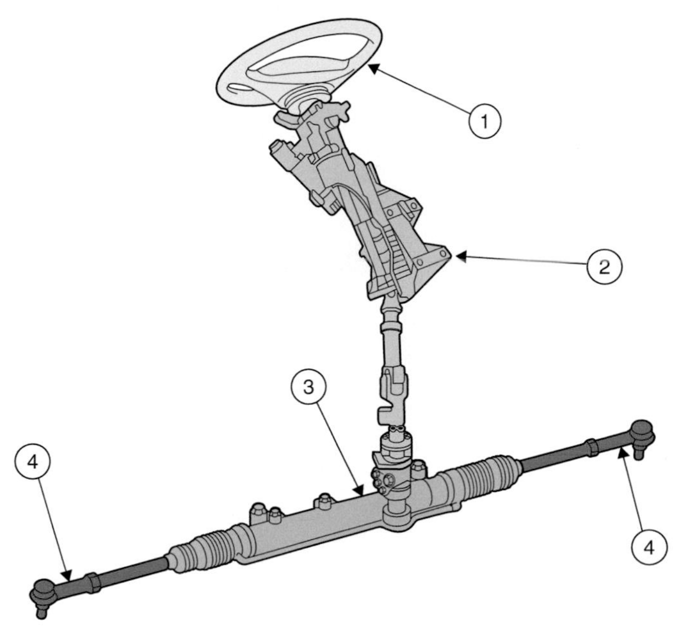

Figure 8.36 Ford steering system

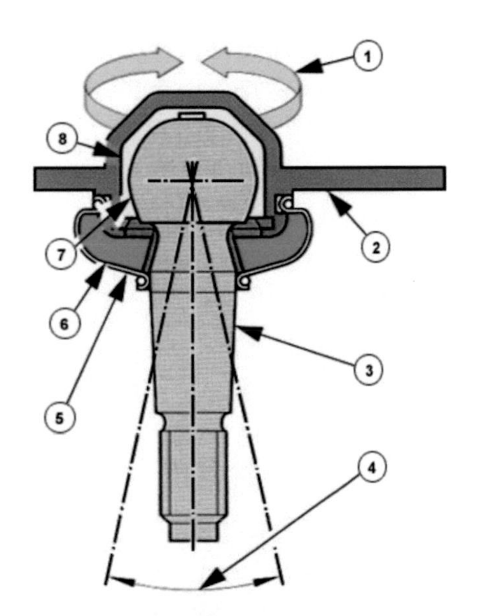

Figure 8.37 Ball joints are used instead of kingpins

Worm and nut steering gear This system consists of a steering screw on which the steering nut is displaced axially as the steering wheel is moved. Slide rings on the circumference of the steering nut transmit the movement to the steering fork and thereby to the drop arm. The drop arm performs a movement of up to 90°. In this type of steering the wear is relatively high. The steering nut play cannot be adjusted, and this is a disadvantage. With this type of steering gear, the steering is linear.

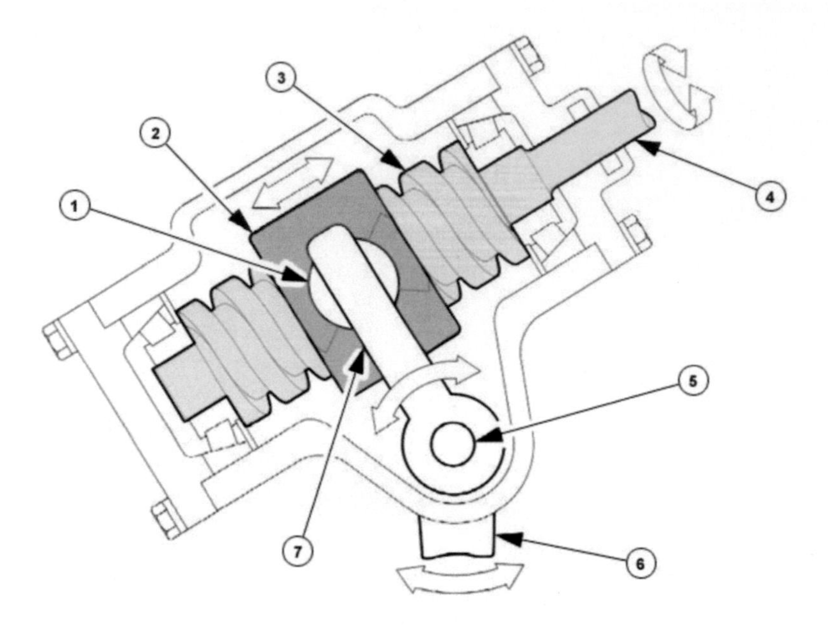

Figure 8.38 Screw and nut steering box

Rack and pinion steering The steering rack housing generally contains a helically toothed pinion, which meshes with the rack. By turning the steering wheel and hence the pinion, the rack is displaced transversely to the direction of travel. A spring-loaded pressure pad presses the rack against the pinion. For this reason the steering gear always functions without backlash. At the same time, the sliding friction between pressure pad and rack acts as a damper to absorb road shocks. Advantages of rack and pinion steering include the shallow construction, the very direct steering, the good steering return and the low cost of manufacture.

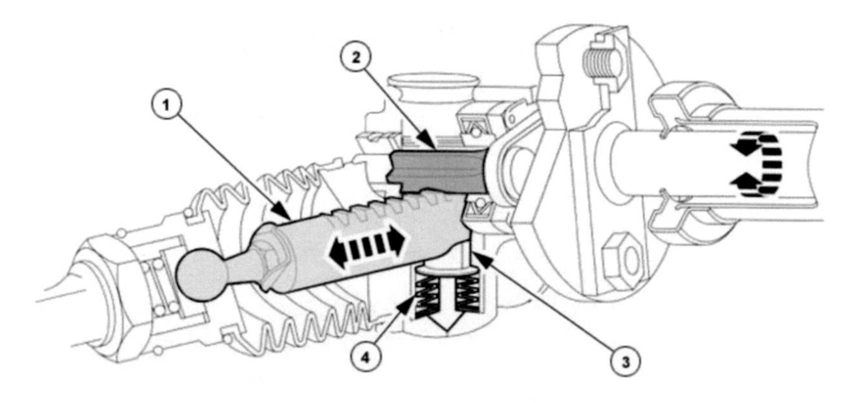

Figure 8.39 Linear steering rack

There is a wide range of steering boxes and steering layouts. On light vehicles, the most common by far, is the steering rack. This is because it has a shallow construction, is very direct, has good steering return and the cost of manufacture is low.

Look back over the previous section and write out a list of the key bullet points.

Make a simple sketch to show how one of the main components or systems in this section operates.

8.1.5 Steering alignment

Measure and adjust tracking Set the arms of the gauges to the wheel size. Zero the tracking gauges by placing the tips of the arms together and adjusting the scale. Roll the vehicle back and forth to make sure the steering is not under stress. Position the gauges to the wheels, take a reading and compare to the manufacturer's specifications.

Figure 8.40 Simple 'mirror' gauges

Tracking (toe in/out) If adjustment is required, check the position of the steering wheel. Adjustment is made by changing the overall length of the track rod. If the spokes are even, make equal adjustments to each end of the track rod. If the spokes are NOT even, turn the wheel until they are and then adjust each end of the track rod, such as to bring the wheels to the correct position.

Track rod adjustment To adjust track rod length, undo the lock nut and then turn the rod, which is threaded into the track rod end. Tighten the lock nuts and check the alignment again. Carry out further adjustment if required – it is quite usual for accurate adjustment to need two or three changes. Secure all components and remove gauges.

Remember! There is a multimedia version of this textbook that includes additional images and interactive features: www.automotivett.com

Figure 8.41 Loosen the lock nut and adjust track rod length

Position the front wheels of the vehicle on turn plates.

Figure 8.42 Car with its wheels on turn plates

Steering geometry Connect measuring equipment to the vehicle as per the manufacturer's instructions. Record readings as follows:

- camber
- castor
- swivel axis/king pin inclination (SAI/KPI)
- offset
- tracking (toe-in/out).

Taking readings The way readings are taken will vary depending on the type of equipment you are using. Lasers are used by some types to allow direct readout from a scale. For different measurements, numerical displays are used.

Figure 8.43 Alignment equipment in use . . .

Figure 8.44 To check castor and SAI

Figure 8.45 LCD readout

Figure 8.46 Laser appears here

Summary Some repairs can involve significant work. However, do not make any compromises. Keep your customers, and yourself, happy and safe.

Look back over the previous section and write out a list of the key bullet points here:

8.2 Routine braking system checks

Together with the multimedia resources, this section is ideal material for students working towards the IMI Awards unit EL09, the City and Guilds 3902–006 unit and all other similar foundation or introduction level qualifications.

After successful completion of this section you will be able to show you have achieved these objectives:

- Be able to work safely.
- Know about braking systems.
- Be able to remove and replace simple brake components and carry out simple checks.
- Know how to dispose of braking system components.

8.2.1 What you must know about braking systems

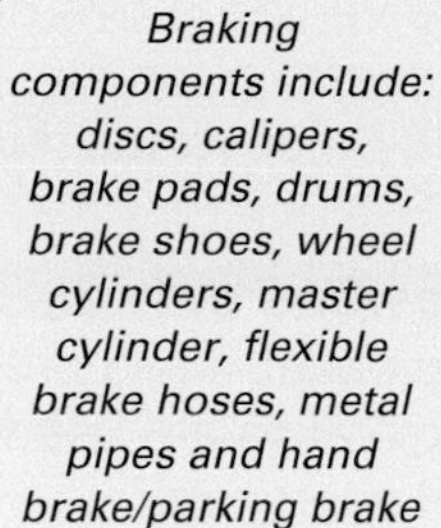

Braking components include: discs, calipers, brake pads, drums, brake shoes, wheel cylinders, master cylinder, flexible brake hoses, metal pipes and hand brake/parking brake

Check and top-up brake fluid reservoir: visually inspect level, check brake fluid type using technical data, clean around cap before removal,

Check operation of brake lights: assistant operate brake lights and chcck brake light operation

How to remove and refit a set of disc pads: jack up vehicle safely, support vehicle with axle stands (if required), remove calliper, remove pads, clean calliper and carrier, clean and lubricate pads where required, refit caliper, check installation after assembly, refit wheel and finally pump brake pedal

How to dispose of brake friction materials (pads and shoes) and brake fluid, and clearing up spillages and disposal of absorbent

8.2.1 Brakes introduction

Energy conversion The main purpose of the braking system is simple; it is to slow down or stop a vehicle. To do this the energy in the vehicle movement must be taken away – or converted. This is achieved by creating friction. The resulting heat takes energy away from the movement. In other words, kinetic energy is converted into heat energy.

Braking system The main braking system of a car works by hydraulics. This means that when the driver presses the brake pedal, liquid pressure forces pistons to apply brakes on each wheel. Disc brakes are used on the front wheels of some cars and on all wheels of sports and performance cars. Braking pressure forces brake pads against both sides of a steel disc. Drum brakes are fitted on the rear wheels of some cars and on all wheels of older vehicles. Braking pressure forces shoes to expand outwards into contact with a drum. The important part of brake pads and shoes is the friction lining.

Brake pads are steel backed blocks of friction material, which are pressed onto both sides of the disc. Older types were asbestos based so you must not inhale the dust. Follow manufacturers' recommended procedures. Pads should be changed when the friction material wears down to 2 or 3 mm. The circular steel disc rotates with the wheel. Some are solid but many have ventilation holes.

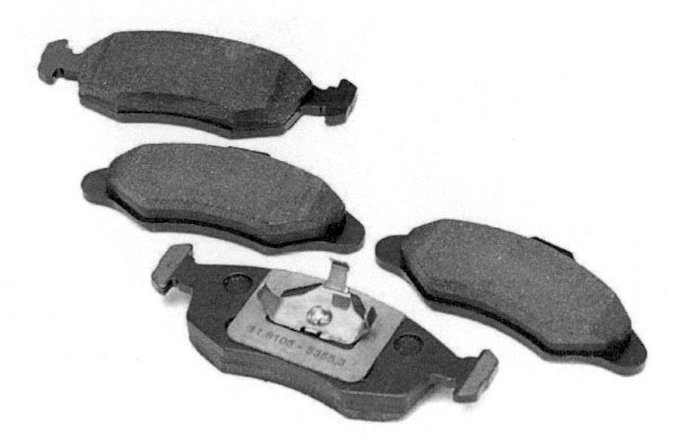

Figure 8.47 Brake pads

Brake shoes Brake shoes are steel crescent shapes with a friction material lining. They are pressed inside a steel drum, which rotates with the wheel. The rotating action of the brake drum tends to pull one brake shoe harder into contact. This is known as self-servo action. It occurs on the brake shoe, which is after the wheel cylinder, in the direction of wheel rotation. This brake shoe is described as the leading shoe. The brake shoe before the wheel cylinder in the direction of wheel rotation is described as the trailing shoe.

Hydraulic cylinders The master cylinder piston is moved by the brake pedal. In its basic form, it is like a pump, which forces brake fluid through the pipes. Pressure in the pipes causes a small movement to operate either brake shoes or pads. The wheel cylinders work like a pump only in reverse.

Figure 8.48 Brake shoes . . .

Figure 8.49 In common use

Brake servo The brake servo increases the force applied by the driver on the pedal. It makes the brakes more effective. Vacuum, from the engine inlet manifold, is used to work most brake servos.

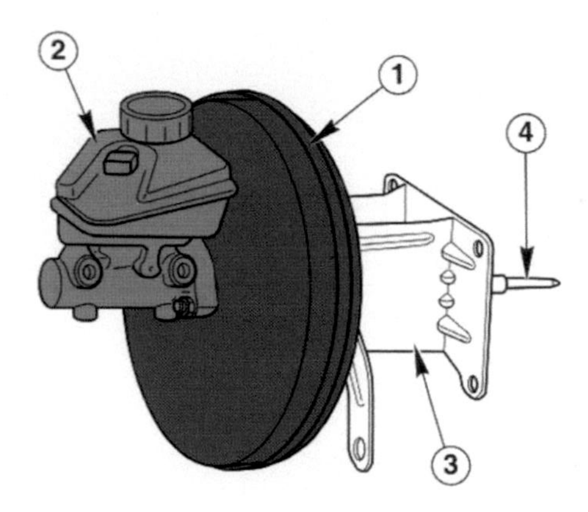

Figure 8.50 Servo unit

Antilock brake system If the brakes cause the wheels to lock and make them skid, steering control is lost. In addition, the brakes will not stop the car as quickly. ABS uses electronic control to prevent this happening.

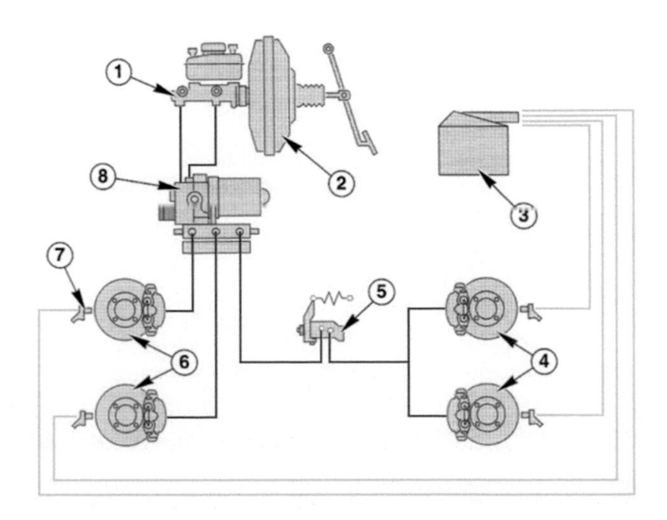

Figure 8.51 ABS layout

Figure 8.52 Flexible pipes

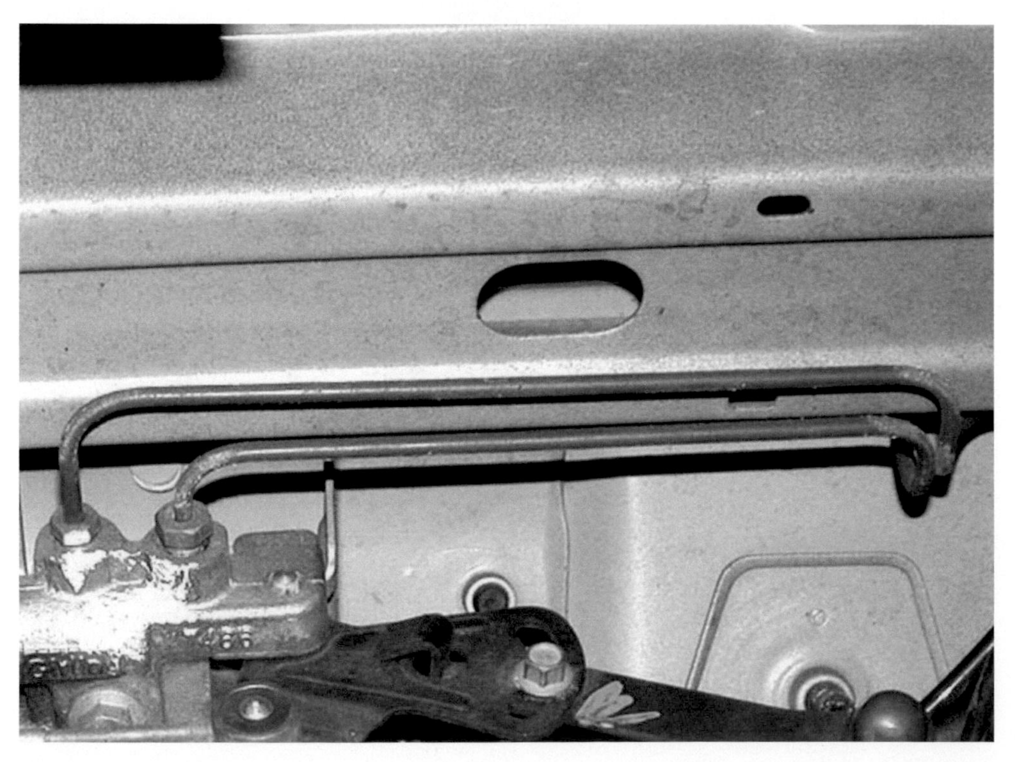

Figure 8.53 Metal pipes

Brake pipes Strong, high-quality pipes are used to connect the master cylinder to the wheel cylinders. Fluid connection, from the vehicle body to the wheels, has to be through flexible pipes to allow suspension and steering movement. As a safety precaution (because brakes are quite important!), brake systems are split into two sections. If one section fails, say, by a pipe breaking, the other will continue to operate.

Test requirements All components of the braking system must be in good working order, in line with most other vehicle systems. Braking efficiency means the braking force compared to the weight of the vehicle. For example, the brakes on a vehicle with a weight of 10 kN (1000 kg x 10 ms–2 [g]) will provide a braking force of, say, 7 kN. This is said to be 70% efficiency. During an annual test, this is measured on brake rollers. The current efficiency requirements in the UK are as follows: service brake efficiency – 50%, second line brake efficiency – 25%, parking brake efficiency – 16%.

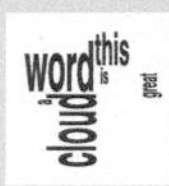

Create a word cloud for one or more of the most important screens or blocks of text in this section.

Look back over the previous section and write out a list of the key bullet points.

8.2.2 Disc, drum and parking brakes

Disc brakes The caliper shown is known as a single acting, sliding caliper. This is because only one cylinder is used but the pads are still pressed equally on both sides of the disc by the sliding action. Disc brakes are less prone to brake fade than drum brakes. This is because they are more exposed and can get rid of heat more easily. They also throw off water better than drum brakes. Brake fade occurs when the brakes become so hot they cannot transfer any more energy – and they stop working!

Disc brakes are self-adjusting. When the pedal is depressed, the rubber seal is pre-loaded. When the pedal is released, the piston is pulled back due to the elasticity of the rubber sealing ring.

Drum brakes Brake shoes are mounted inside a cast iron drum. They are mounted on a steel backplate, which is rigidly fixed to a stationary part of the axle. The two curved shoes have friction material on their outer faces. One end of each shoe bears on a pivot point. The other end of each shoe is pushed out by the action of a wheel cylinder when the brake pedal is pressed. This puts the brake linings in contact with the drum inner surface. When the brake pedal is released, the return spring pulls the shoes back to their rest position.

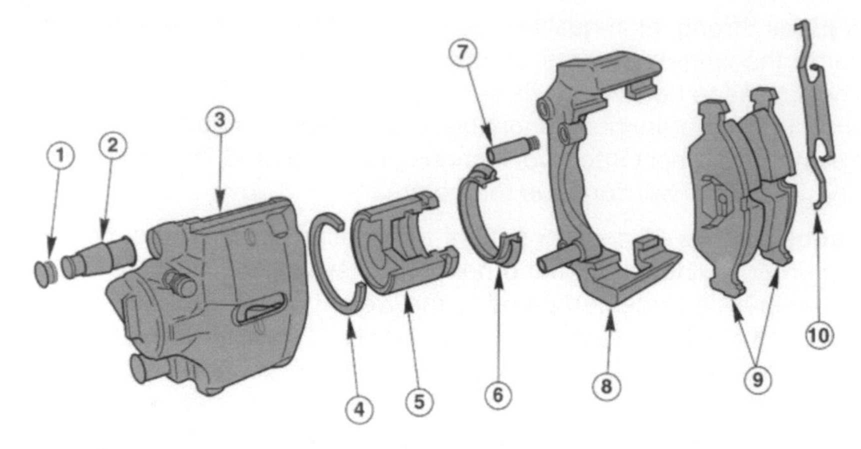

Figure 8.54 Sliding disc brake caliper components

Figure 8.55 Sliding disc brake caliper

Drum brake features Drum brakes are more adversely affected by wet and heat than disc brakes, because both water and heat are trapped inside the drum. However, they are easier to fit with a mechanical hand brake linkage.

Brake adjustments Brakes must be adjusted so that the minimum movement of the pedal starts to apply the brakes. The adjustment in question is the gap between the pads and disc and the shoes and drum. Disc brakes are self-adjusting because as pressure is released it moves the pads just away from the disc. Drum brakes are different because the shoes are moved away from the drum to a set position by a pull off spring. Self-adjusting drum brakes are almost universal now on light vehicles. A common type uses an offset ratchet, which clicks to a wider position if the shoes move beyond a certain amount when operated.

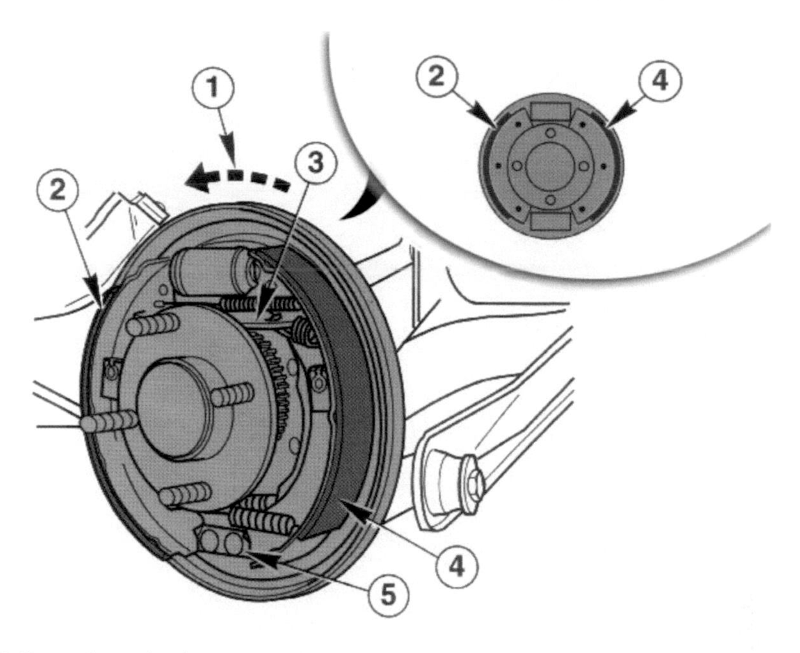

Figure 8.56 Rear drum brake

Figure 8.57 Brake drum

Manual adjustment Adjustment through a hole in the back plate is often used. This involves moving a type of nut on a threaded bar, which pushes the shoes out as it is screwed along the thread. This method is similar to the automatic adjusters. An adjustment screw on the back plate is now quite an old method. A screw or square head protruding from the back plate moves the shoes by a snail cam. As a guide, tighten the adjuster until the wheels

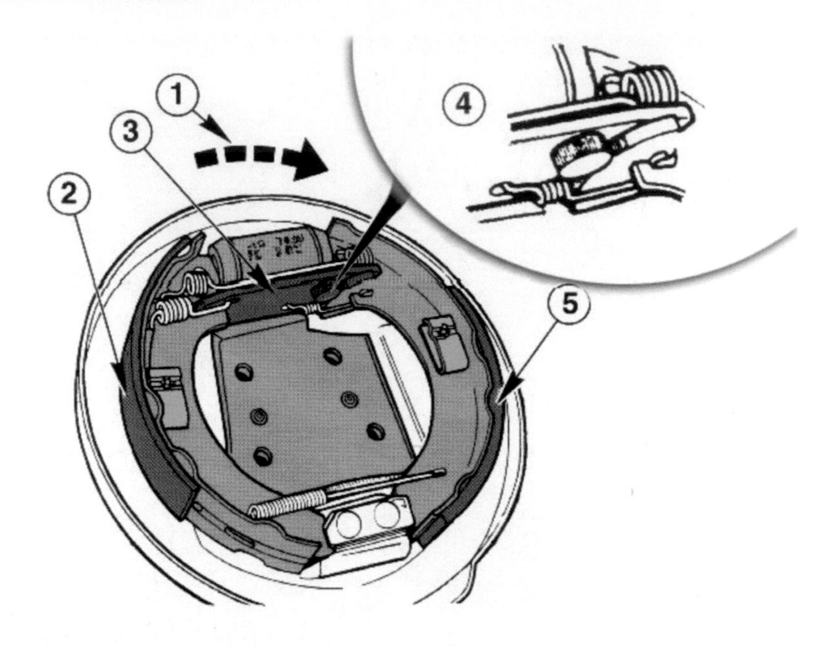

Figure 8.58 Self-adjusting device

Figure 8.59 Brake adjustment hole

lock, and then move it back until the wheel is just released. You must ensure that the brakes are not rubbing as this would build up heat and wear the friction material very quickly.

The precise way in which the shoes move into contact with the drum affects the power of the brakes. If the shoes are both hinged at the same point then the system is said to have one leading and one trailing shoe. As the shoes are pushed into contact with the drum, the leading shoe is dragged by the drum rotation harder into contact, whereas the rotation tends to push the trailing shoe away. This 'self-servo' action on the leading shoe can be used to

Figure 8.60 Square-type adjuster (old method)

increase the power of drum brakes. This is required on the front wheels of all-round drum brake vehicles.

Inside a brake drum, the **hand brake linkage** is usually a lever mechanism as shown here. This lever pushes the shoes against the drum and locks the wheel. The hand brake lever pulls on one or more cables and has a ratchet to allow it to be locked in the on position. There are a number of ways in which the hand brake linkage can be laid out to provide equal force, or compensation, for both wheels:

- Two cables, one to each wheel.
- Equaliser on a single cable pulling a 'U' section to balance effort through the rear cable (as shown here).
- Single cable to a small linkage on the rear axle.

Disc-type handbrake Some sliding caliper disc brakes incorporate a handbrake mechanism. The footbrake operates as normal. Handbrake operation is by a moving lever. The lever acts through a shaft and cam, which works on the adjusting screw of the piston. The piston presses one pad against the disc and because of the sliding action, the other pad also moves.

Some manufacturers use a set of small brake shoes inside a small drum, which is built in to the brake disc. The caliper is operated as normal by the footbrake. The small shoes are moved by a cable and lever.

In **summary**, remember that the purpose of the braking system is to slow down or stop a vehicle. This is achieved by converting the vehicle's movement energy into heat. Friction is used to do this. Braking system developments have improved efficiency, reliability and ease of servicing.

Remember! There is a multimedia version of this textbook that includes additional images and interactive features: www.automotivett.org

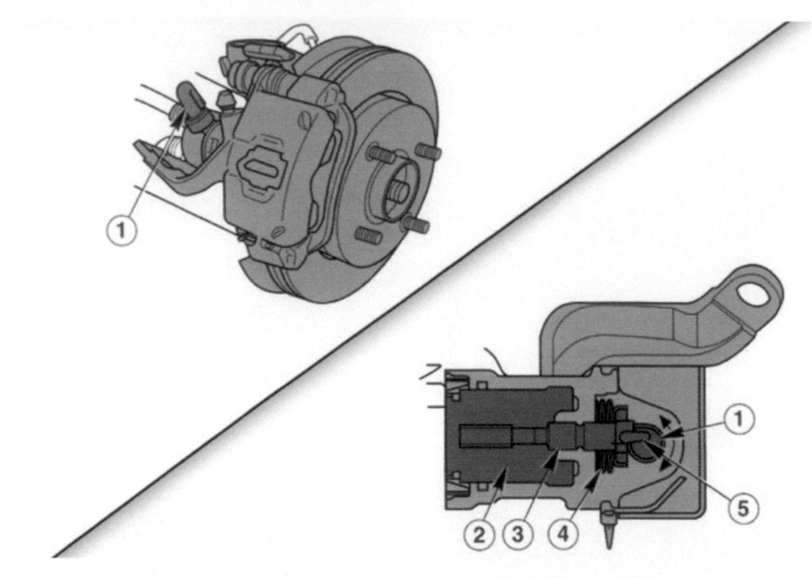

Figure 8.61 Sliding caliper parking brake

Look back over the previous section and write out a list of the key bullet points.

Answer the following questions either here in your book or electronically.

1 State why disc brakes are self-adjusting.
2 Describe the operation of a handbrake

8.3 Routine wheel and tyre checks

Together with the multimedia resources, this section is ideal material for students working towards the IMI Awards unit EL10, the City and Guilds 3902–009 unit and all other similar foundation or introduction level qualifications.

After successful completion of this section you will be able to show you have achieved these objectives:

- Be able to work safely.
- Know how wheels and tyres are constructed.
- Know wheel and tyre terminology.
- Be able to safely and correctly remove and refit road wheels.
- Be able to check tyre pressure and tread depth.

8.3.1 What you must know about wheels and tyres

The correct sequence and procedure for removing and refitting a wheel:

1. positioning of vehicle and jack points
2. operation and positioning of jack
3. location of axle stands
4. loosening wheel nuts
5. removal of wheel
6. refitting wheel
7. sequence for tightening wheel nuts
8. wheel nut torque setting
9. use of torque wrench and wheel brace
10. removal of axle stands
11. lowering vehicle
12. correct storage of tools and equipment
13. check vehicle and removal of grease marks

The common types of tyre used on light vehicles and their construction details: radial ply, cross ply, tubed and tubeless tyres. The arrangement of the casing plies, tyre tread, depth indicator, bead, tread bracing, sidewall and tyre markings

The main markings and terminology associated with vehicle wheels and tyres to include: tyre type, tyre size, tread depth, speed rating and wheel diameter

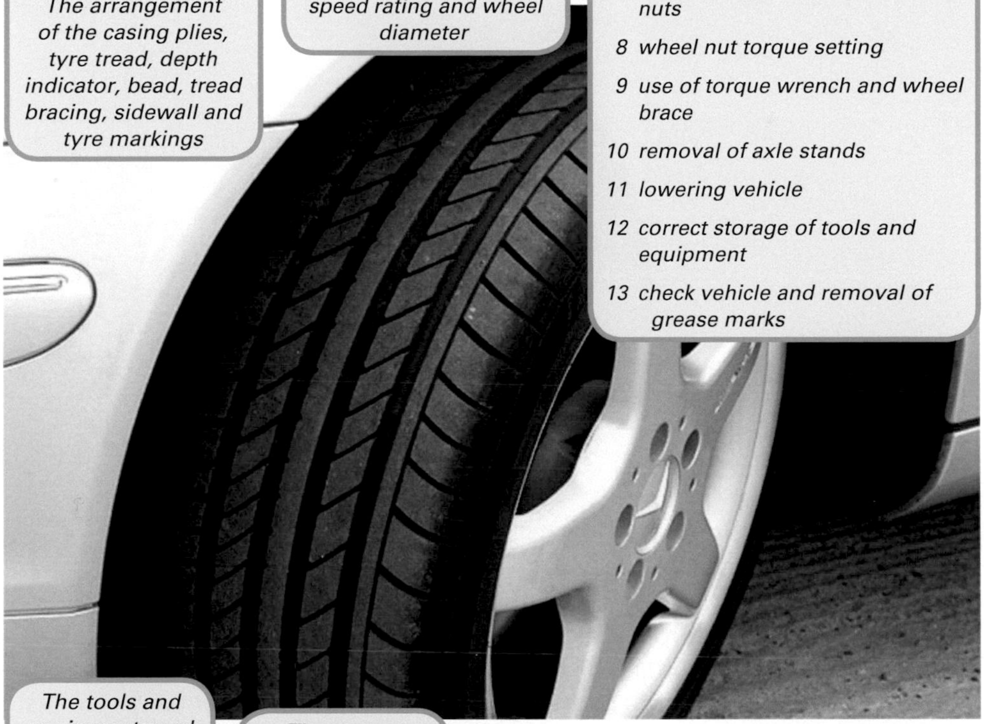

The tools and equipment used for checking and inspecting wheels and tyres: tread depth indicator, tyre inflator, pressure gauge, torch, light hammer, axle stands, wheel chocks, car

The correct procedure for checking and inspecting wheels and tyres: tyre pressure, tread depth, excessive tyre wear, cuts and bulges, wheel damage, buckled wheels, security of wheels, and tightness of wheel nuts

The common types of wheel used on light vehicles to include: alloy wheels, pressed steel wheels, wire wheels and space saver

Always wear suitable PPE and use tools and equipment safely and correctly when doing practical work on any vehicle

8

8.3.2 Wheels

Introduction Together with the tyre, a road wheel must support the weight of the vehicle. It must also be capable of withstanding a number of side thrusts when cornering, and torsional forces when driving. Road wheels must be strong, but light weight. They must be cheap to produce, easy to clean, and simple to remove and refit.

Spoked wheels Spoked wheels are attractive but tend only to be used on older sports cars. They are a smaller diameter, but stronger version, of a bike wheel. These wheels must have tyres with an inner tube. Spoked wheels allow good ventilation and cooling for the brakes but can be difficult to keep clean!

Pressed steel wheels The centre of this type of wheel is made by pressing a disc into a dish shape, to give it greater strength. The rim is a rolled section, which is circled and welded. The rim is normally welded to the flange of the centre disc. The centre disc has a number of slots under the rim. This is to allow ventilation for the brakes as well as the wheel itself.

Figure 8.62 Standard wheel used on many cars

Steel wheel rim features The manufacture of this type of wheel makes it cheap to produce and strong. The bead of a tyre is made from wire, which cannot be stretched for fitting or removal. The wheel rim therefore, must be designed to allow the tyre to be held in place, but also allow for easy removal.

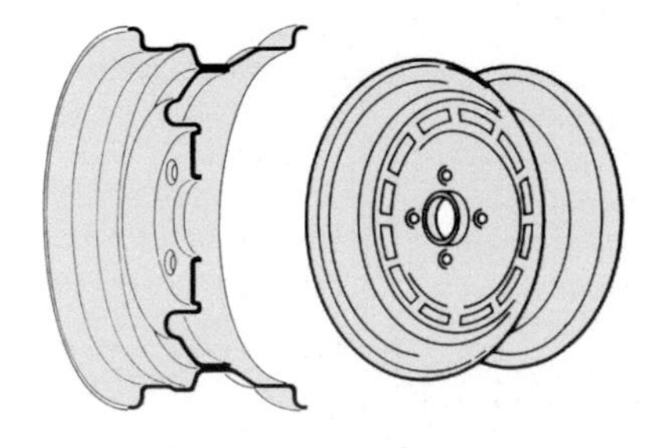

Figure 8.63 Standard wheel design

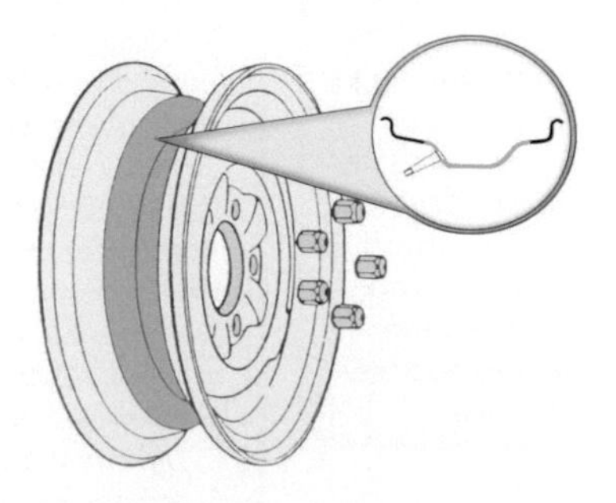

Figure 8.64 Wheel profile

'Well-base' wheel To facilitate fitting and removal a 'well-base' is manufactured into the rim. For tyre removal, one bead must be forced into the well. This then allows the other bead to be levered over the edge of the rim. The bead seats are made with a taper so that as the tyre is inflated the bead is forced up the taper by the air pressure. This locks the tyre on to the rim making a good seal.

Wheel trims Steel wheels are a very popular design. They are very strong and cheap to produce. Steel wheels are usually covered with plastic wheel trims. Trims are available in many different styles.

Figure 8.65 Wheel trim

Alloy wheels Alloy wheels, or 'alloys', are good, attractive looking wheels. They tend to be fitted to higher specification vehicles. Many designs are used. They are light weight but can be difficult to clean.

Figure 8.66 Alloy wheels

Cast alloy wheels A large number of vehicles are fitted with wheels made from alloy. Wheels of this type are generally produced from aluminium alloy castings, which are then machine finished. Alloy wheels can be easily damaged by 'kerbing'!

Advantages of alloy wheels The main advantage of cast alloy road wheels is their reduced weight, and of course, they look good. Disadvantages are their lower resistance to corrosion, and that they are more prone to accidental damage. The general shape of the wheel, as far as tyre fitting is concerned, is much the same as the pressed steel type.

Look back over the previous section and write out a list of the key bullet points.

Use the interactive media search tools to look for pictures and videos to examine the subject in this section in more detail.

Fixing the road wheels Light vehicle road wheels are usually held in place by four nuts or bolts. The fixing holes in the wheels are stamped or machined to form a cone-shaped seat.

Figure 8.67 Ten . . .

Wheel nuts and bolts The wheel nut or bolt heads, fit into this seat. This ensures that the wheel fits in exactly the right position. In the case of the steel pressed wheels, it also strengthens the wheel centre round the stud holes.

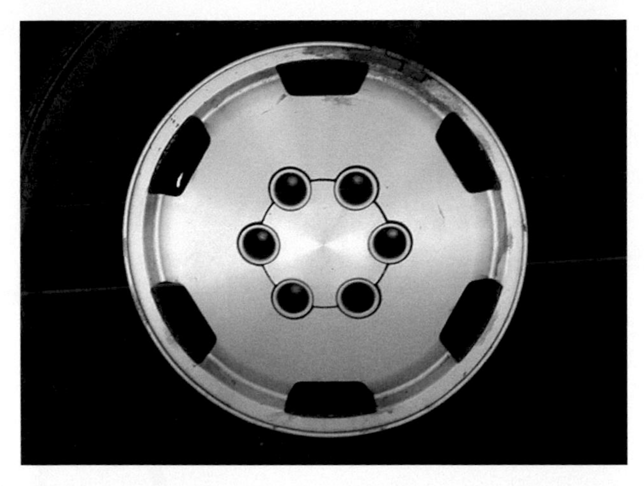

Figure 8.68 Six . . .

Figure 8.69 Five . . .

Figure 8.70 Four stud fixings

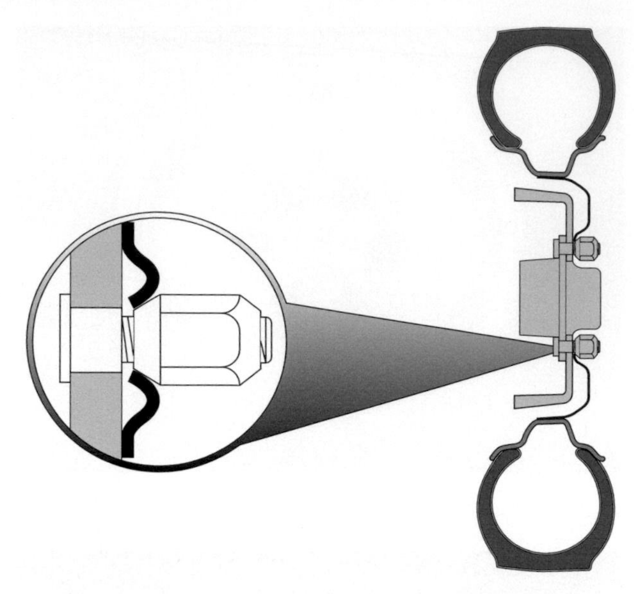

Figure 8.71 Wheel fixing

Fitting a wheel When fitting a wheel, the nuts or bolts must be tightened evenly in a diagonal sequence. It is also vital that they are set to the correct torque. Ensure the cone shaped end of the wheel nuts is fitted towards the wheel.

Wheel rim measurement Car wheel rim measurement consists of three main dimensions as shown. The nominal rim diameter is the distance

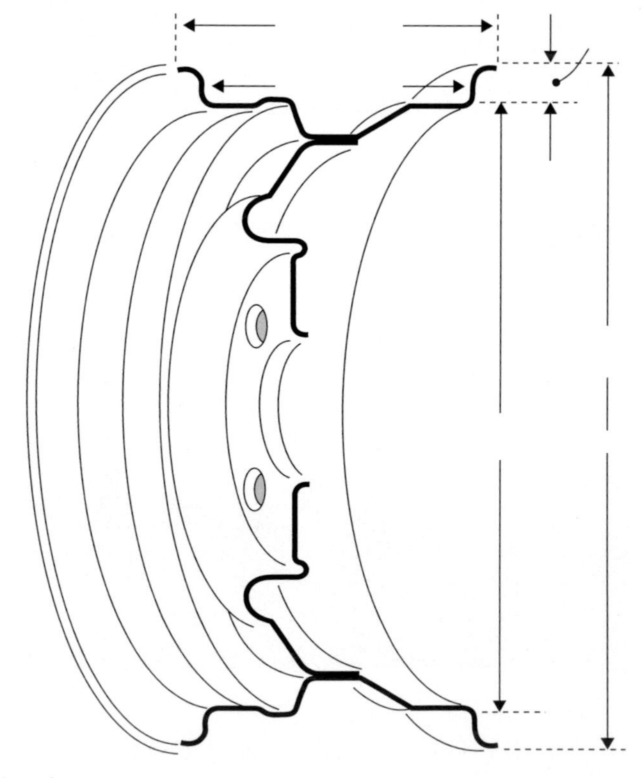

Figure 8.72 Rim sizes

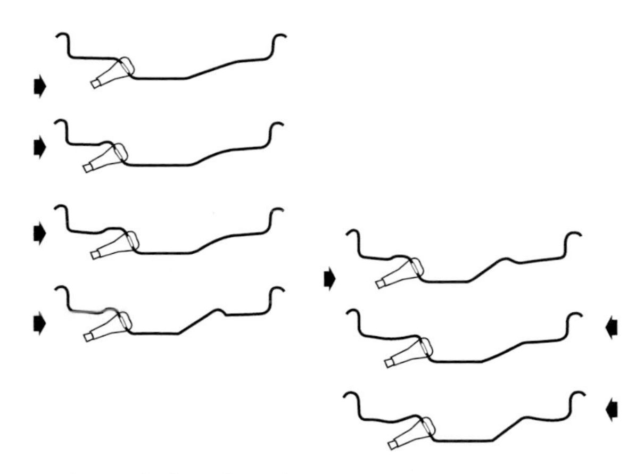

Figure 8.73 Rims and tyre fitting directions

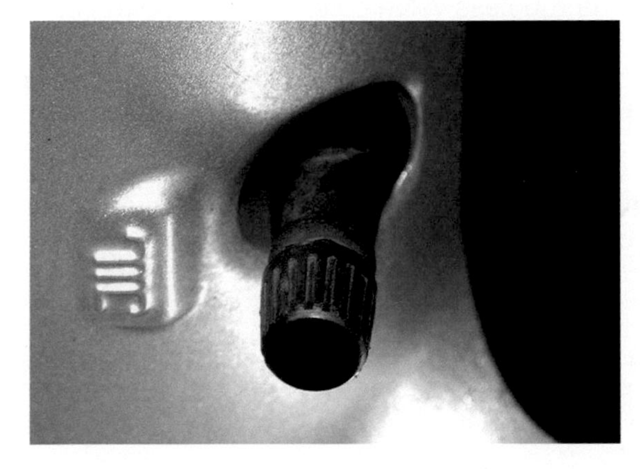

Figure 8.74 Valve on a wheel

between the bead seats. The inside rim width, is the distance between rims. It is not possible to measure this when the tyre is fitted. The flange height can be determined by subtracting the nominal diameter, from the outside rim diameter.

Types of rims There are many types of car wheel rims. The picture shows a selection of those in common use. The arrows indicate the side of the rim over which the tyre should be removed or fitted.

Valves The valve is to allow the tyre to be inflated with air under pressure, prevent air from escaping after inflation, and to allow the release of air for adjustment of pressure. The valve assembly is contained in a brass tube, which is bonded into a rubber sleeve and mounting section.

Valve core The valve core consists of a centre pin, which has metal and rubber disc valves. When the tyre is inflated, the centre pin is depressed, the disc valve moves away from the bottom of the seal tube and allows air to enter the tyre. To release air, or for pressure checking, the centre pin is depressed. During normal operation, the disc valve is held onto its seat by a spring and by the pressure of air. If all the air needs to be released, the valve core assembly can be removed. The upper part of the valve tube is threaded to accept a valve cap. This prevents dirt and grit from entering and acts as a secondary seal.

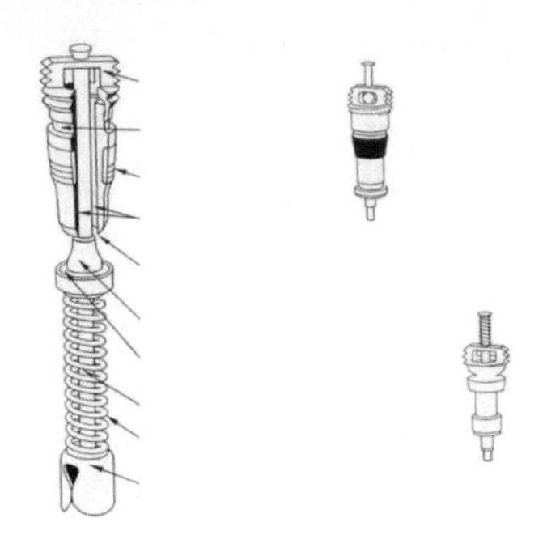

Figure 8.75 Details of the valve construction

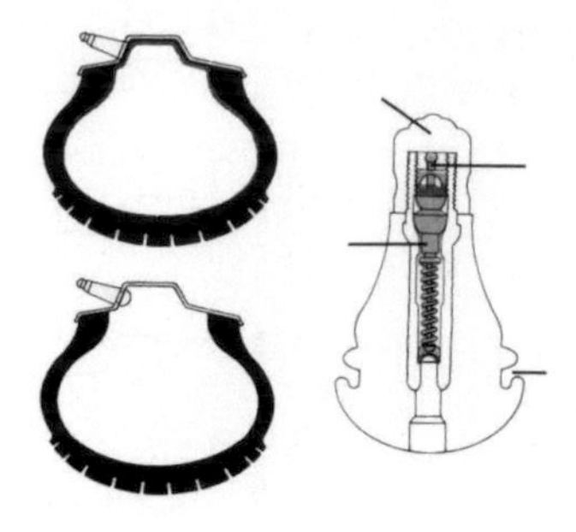

Figure 8.76 Tyres and valves

Tubeless valve The tubeless valve core is as described previously. However, the valve body must be made so that when fitted into the wheel, an airtight seal is formed. Wheel rims used for tubeless tyres must be sealed and airtight. Most wheels and tyres in use are of the tubeless design.

Look back over the previous section and write out a list of the key bullet points.

Make a simple sketch to show how one of the main components or systems in this section operates.

8.3.3 Tyres

Basic functions The tyre performs two basic functions. It acts as the primary suspension, cushioning the vehicle from the effects of a rough surface. It also provides frictional contact with the road surface. This allows the driving wheels to move the vehicle. The tyres allow the front wheels to steer and the brakes to slow or stop the vehicle.

Pneumatic tyres The tyre is a flexible casing, which contains air. Tyres are manufactured from reinforced synthetic rubber. The tyre is made from an inner layer of fabric plies, which are wrapped around bead wires at the inner

Figure 8.77 Pneumatic tyre construction

edges. The bead wires hold the tyre in position on the wheel rim. The fabric plies are coated with rubber, which is moulded to form the side walls and the tread of the tyre. Behind the tread is a reinforcing band, usually made of steel, rayon or glass fibre. Modern tyres are mostly tubeless, so they also have a thin layer of rubber coating the inside to act as a seal.

Radial tyre carcass An innermost sheet of airtight synthetic rubber performs the 'inner tube' function. The carcass ply is made up of thin textile fibre cables, laid out in straight lines and bonded into the rubber. These cables are largely responsible for determining the strength of the tyre structure. The carcass ply of a car tyre has about 1,400 cables, each capable of withstanding 15 kg. A lower filler is responsible for transferring propulsion and braking torques, from the wheel rim, to the road surface.

Figure 8.78 Tyre carcass construction

Figure 8.79 Radial tyre make-up

Radial tyre features Beads clamp the tyre firmly against the wheel rim. The beads can withstand forces up to 1,800 kg. The tyre has supple rubber walls, which protect the tyre against impacts (with kerbs, etc.) that might otherwise damage the carcass. There is also a hard rubber link between the tyre and the rim. Crown plies, consist of oblique overlapping layers of rubber reinforced with very thin, but very strong, metal wires. The overlap between these wires, and the carcass cables, forms a series of non-deformable triangles. This arrangement lends great rigidity to the tyre structure.

Tyre markings Markings on the sides of tyres are quite considerable and can be a little confusing. The following is a list of the information given on modern tyres. The size, speed and load headings will be examined in more detail.

- size (e.g. 195/55–15)
- speed rating (e.g. H, V, Z)
- load index (e.g. 84, 89, 92,)
- UTQG ratings (temperature, traction, tread wear)
- M&S designation
- maximum load
- maximum pressure
- type of construction
- EU approval mark
- US approval mark
- manufacture date.

Tyre Sizes A tyre's size is expressed in the format WWW/AA-DD (e.g. 195/55–15).

WWW is the tyre's sidewall-to-sidewall width in millimetres (195). AA is the aspect ratio or profile (55). This gives the tyre's height as a percentage of its width. DD is the diameter of the wheel in inches (15). Some tyres now also give this in millimetres. If the size is shown as P195/55R15, the 'P' stands for passenger and the 'R' is for radial ply construction.

Figure 8.80 Side of a tyre showing markings

Figure 8.81 Side of a tyre showing markings

Older tyres For an older tyre without an aspect ratio (e.g. 195R13), it is assumed to be about an 80 series tyre (195/80R13). The practice of listing the aspect ratio is now more common. The speed rating was traditionally shown as a part of the tyre's size (e.g. 195/55VR15). Since the inclusion of load ratings, many manufacturers now show the speed rating after the size in combination with the load rating (e.g. 195/55R15 84V).

Speed ratings Commonly used speed ratings are shown in the following table.

Load index The load index indicates the maximum weight the tyre can carry at the maximum speed indicated by its speed rating. Some 'Load Rating Indices' are listed in the following table.

Rating	Certified maximum speed (km/h)	Certified maximum speed (mph)
N	140	88
Q	160	100
S	180	112
T	190	118
U	200	124
H	210	130
V	240	150*
Z	Over 240	Over 150
W	270	169
Y	300	188

*Note that originally, V was 'over 130mph'. As W and Y ratings are now used, Z is redundant.

Performance tyre tread The type of tyre tread shown here uses a directional pattern for improved water evacuation. The centre rib gives improved steering control. Overall, this tyre, manufactured by Avon, gives improved wet grip and reduced noise.

Rating	Capacity(kg)	Capacity(lb.)
75	387	853
82	475	1047
84	500	1102
85	515	1135
87	545	1201
88	560	1235
91	615	1356
92	630	1389
93	650	1433
105	925	2030

Standard tyre tread This tyre is built for the cost conscious motorist. It saves fuel when fitted all round because of a lower rolling resistance. The rubber compound used, prolongs tyre life. It has been made using environmentally neutral methods.

Remember! There is a multimedia version of this textbook that includes additional images and interactive features: www.automotivett.org

Figure 8.82 Directional pattern tread

Figure 8.83 Avon 'cost conscious' tyre

Look back over the previous section and write out a list of the key bullet points.

Create a mind map to illustrate the important features of a component or system in this section.

8.4 Chassis systems puzzles

8.4.1 Cryptograms

A	B	C	D	E	F	G	H	I	J	K	L	M	N	O	P	Q	R	S	T	U	V	W	X	Y	Z
																		24		19					

													S			S	U	S				S		
18	11	23	12	3	6	7	23		26	23	23	13	24		18	24	24	24	17	23	26	24	3	21	26

S		S													U							S					S
24	22	24		1	23	16		1	21		13	18	16	17		21	19	1		10	21	18	13		24	12	21	6	2	24

Cryptogram 8.1 Smooth ride:

8.4.2 Anagrams

Noises Spun   Session Pun
Poises Nuns   Pisses Noun

Anagrams 8.1 You can be kept in this: what word did I use to get these anagrams?

8.4.3 Word search

```
N Y S P C N I Q K L D S R G N
H O H H U A P S D E R J F K N
M P I V A H H M R E C R U T O
Z G H S V C S H B E I D F V I
N C Y T N E K R Q C P H U J S
J P D N M E O L T O O M N I R    ABSORBERS
T Y R C T S P I E E C U A L O    DAMPERS
K M O G B D O S H S S H S D T    FRICTION
E E G A S N J Z U Q E U D W T    HYDROGAS
S W A S G N I R P S L I O W R    SHACKLES
W I S H B O N E E O E K N L M    SPRINGS
V G T F X Q Y B D R T V M X N    SUSPENSION
H P I A P J F D B S G F U E J    TELESCOPIC
O S P Q M C E Z J X S Q Y U C    TORSION
I F C F G A E S V K A M Y Y A    WISHBONE
```

Word search 8.1 Suspension

8.4.4 Crosswords

Across

1 The hydraulic brake component inside a brake drum that moves the shoes (5,8)

3 A special liquid used in hydraulic systems to operate in slave cylinders or calipers (5,5)

4 Parking mechanism to stop the car rolling away

5 Rotating item inside which brake shoes operate

6 Reduction of braking effort caused by overheating (5,4)

9 A steel plate with a bonded friction lining used on disc brakes

10 Creates pressure in a hydraulic brake system by a piston in a tube (6,8)

11 A system for transmitting and modifying force by the use of a liquid

14 Instrument used to check moisture content of brake fluid

15 Inclination of a wheel compared to the vertical

Down

2 ABS (8,7,6)

7 The act of removing air from a liquid system

8 In a disc brake system, the mechanism that squeezes brake pads on to the disc

12 Part of drum brake carrying a friction linings

13 A mechanism that increases the braking effort applied by the driver

Crossword 8.1 Brakers

CHAPTER 9

Transmission systems

9.1 L111 Introduction to Vehicle Transmission Systems

Together with the multimedia resources, this section is ideal material for students working towards the IMI Awards unit L111, the City and Guilds 3902–007 unit and all other similar foundation or introduction level qualifications.

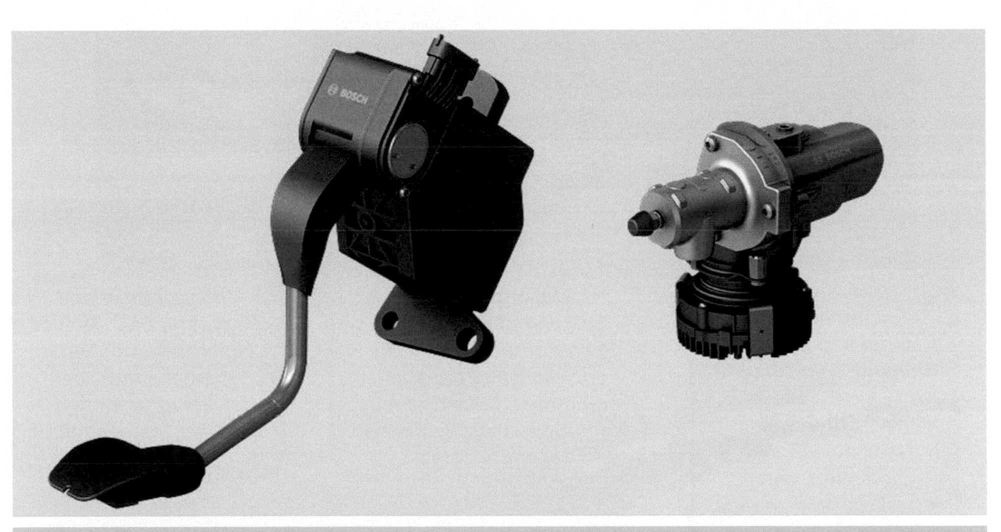

After successful completion of this section you will be able to show you have achieved these objectives:

- Be able to work safely.
- Know vehicle transmission drivelines.
- Know about vehicle gearboxes.
- Be able to carry out routine maintenance on vehicle drivelines.

Automotive Technician Training: Entry Level 3. 978-0-415-72040-3.
© Tom Denton. Published by Taylor & Francis. All rights reserved.

9.1.1 What you must know about transmission systems

Always wear suitable PPE and use tools and equipment safely and correctly when doing practical work on any vehicle

The main components of a manual gearbox and clutch: clutch plate, cover assembly, thrust bearing, casing, gears and selectors

The main automatic transmission and torque converter components: torque converter/ fluid flywheel, housing, multiplate clutches, brake bands, hydraulic control and epicyclic gears

The purpose of the gearbox, provide permanent neutral, reverse gear, increase torque, allow vehicle to accelerate and allow vehicle to reach suitable top speed

Check and top up lubricant level/s in a manual and automatic transmission system: vehicle on level surface, prerequisites for checking level followed, level checked, refer to data for correct lubricant used to top up

The removal and refitting of a driveshaft gaiter or propshaft joint: correct use of tools and equipment, bolts correctly tightened, correctly aligned and lubricated

The assembly and alignment of a manual clutch assembly onto an engine flywheel: cleaning of friction surfaces, installation of friction plate

The vehicle transmission layouts: engine, clutch, gearbox, driveshaft/propshaft, final drive, centre differential/ transfer box

9.1.2 Clutch

A **clutch** is a device for disconnecting and connecting rotating shafts. In a vehicle with a manual gearbox, the driver depresses the clutch when changing gear, thus disconnecting the engine from the gearbox. It allows a temporary neutral position for gear changes and also a gradual way of taking up drive from rest.

The driver operates the clutch by pushing down a pedal. This movement has to be transferred to the release mechanism. There are two main methods used. These are cable and hydraulic. The cable method is the most common. Developments are taking place and an electrically operated clutch will soon be readily available.

Cable A steel cable is used, which runs inside a plastic coated steel tube. The cable 'outer' must be fixed at each end. The cable 'inner' transfers the movement. One problem with cable clutches is that movement of the engine, with respect to the vehicle body, can cause the length to change. This results in a judder when the clutch is used. This problem has been almost eliminated, however, by careful positioning and quality engine mountings.

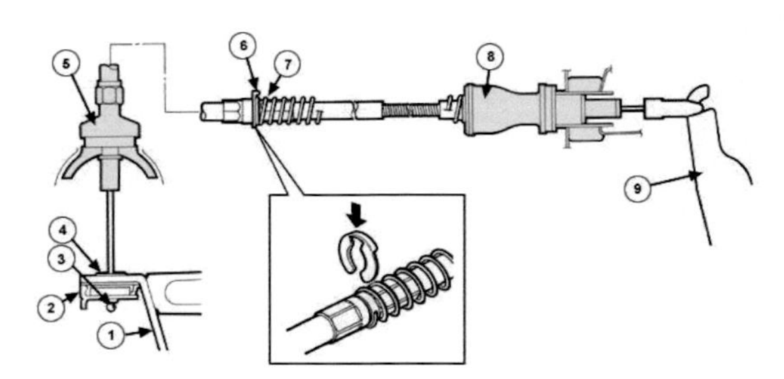

Figure 9.1 Clutch cable

This clutch cable works on a simple lever principle. The clutch pedal is the first lever. Movement is transferred from the pedal to the second lever, which is the release fork. The fork in turn, moves the release bearing to operate the clutch.

Hydraulic A hydraulic mechanism involves two cylinders. These are termed the master and slave cylinders. The master cylinder is connected to the clutch pedal. The slave cylinder is connected to the release lever.

The clutch pedal moves the master cylinder piston. This pushes fluid through a pipe, which in turn forces a piston out of the slave cylinder. The movement ratio can be set by the cylinder diameters and the lever ratios.

Make a simple sketch to show how one of the main components or systems in this section operates.

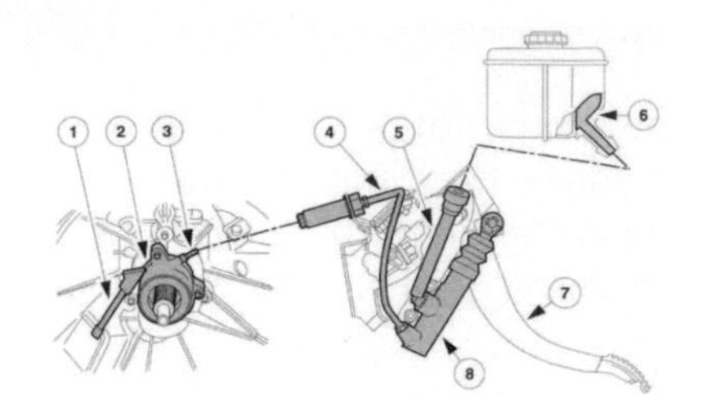

Figure 9.2 Clutch hydraulic components

Basic functions A clutch is a device for disconnecting and connecting rotating shafts. In a vehicle with a manual gearbox, the driver pushes down the clutch when changing gear to disconnect the engine from the gearbox. It also allows a temporary neutral position for, say, waiting at traffic lights and a gradual way of taking up drive from rest.

Figure 9.3 Diaphragm clutch

Clutch location The exact location of the clutch varies with vehicle design. However, the clutch is always fitted between the engine and the transmission. With few exceptions, the clutch and flywheel are bolted to the rear of the engine crankshaft.

Main parts The clutch is made of two main parts, a pressure plate and a driven plate. The driven plate, often termed the clutch disc, is fitted on the shaft, which takes the drive into the gearbox.

Figure 9.4 Driven plate and pressure plate together with bearings kit and an alignment tool (Source: http://haysclutches.com)

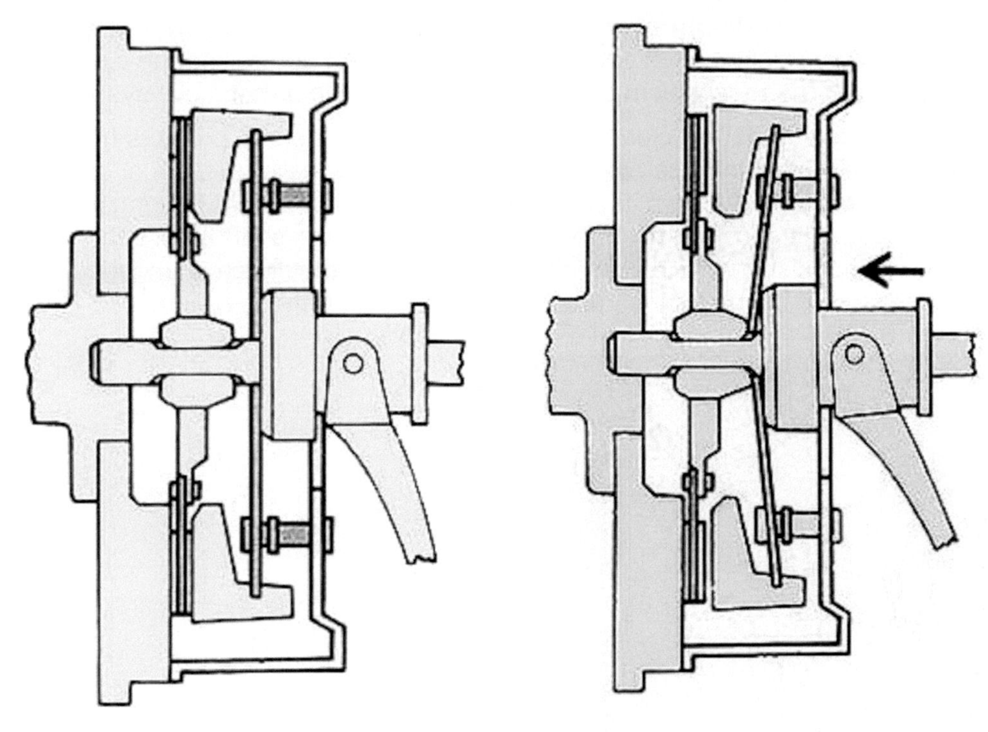

Figure 9.5 Clutch engaged

Figure 9.6 Clutch disengaged

Engagement When the clutch is engaged, the pressure plate presses the driven plate against the engine flywheel. This allows drive to be passed to the gearbox. Depressing the clutch moves the pressure plate away, which frees the driven plate.

Coil springs Earlier clutches, and some heavy-duty types, use coil springs instead of a diaphragm. However, the diaphragm clutch replaced the coil spring type because it has the following advantages:

- It is not affected by high speeds (coil springs can be thrown outwards).
- The low pedal force makes for easy operation.
- It is light and compact.
- The clamping force increases or at least remains constant as the friction lining on the plate wears.

Figure 9.7 Coil spring clutch assembly

Movement of the diaphragm clutch The animation shows the movement of the diaphragm during clutch operation. The method of controlling the clutch is quite simple. The mechanism consists of either a cable or hydraulic system.

Clutch shaft The clutch shaft, or gearbox input shaft, projects from the front of the gearbox. Most shafts have a smaller section or spigot, which projects from its outer end. This rides in a spigot bearing in the engine crankshaft flange. The splined area of the shaft allows the clutch disc to move along the splines. When the clutch is engaged, the disc drives the gearbox input shaft through these splines.

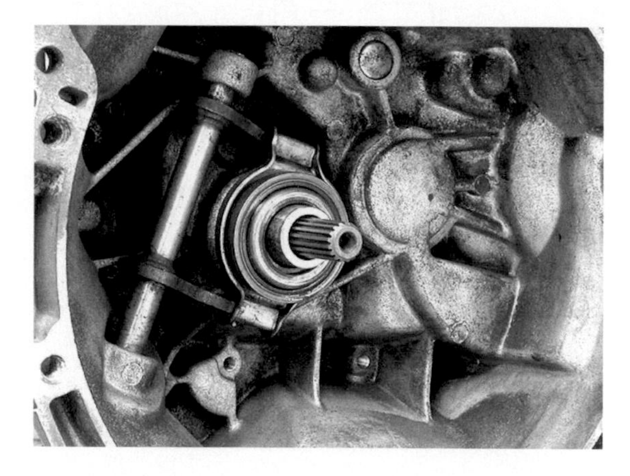

Figure 9.8 Gearbox input shaft

Figure 9.9 Clutch disc or driven plate

Clutch disc The clutch disc is a steel plate covered with frictional material. It fits between the flywheel face and the pressure plate. In the centre of the disc is the hub, which is splined to fit over the splines of the input shaft. As the clutch is engaged, the disc is firmly squeezed between the flywheel and pressure plate. Power from the engine is transmitted by the hub to the gearbox input shaft. The width of the hub prevents the disc from rocking on the shaft as it moves along the shaft.

Frictional facings The clutch disc has frictional material riveted or bonded on both sides. These frictional facings are either woven or moulded. Moulded facings are preferred because they can withstand high pressure plate loading forces. Grooves are cut across the face of the

Figure 9.10 Friction material

friction facings to allow for smooth clutch action and increased cooling. The cuts also make a place for the facing dust to go as the clutch lining material wears.

Health hazards The frictional material wears as the clutch is engaged. At one time asbestos was in common use. Due to awareness of the health hazards resulting from asbestos, new lining materials have been developed. The most commonly used types are paper-based and ceramic materials. They are strengthened by the addition of cotton and brass particles, and wire. These additives increase the torsional strength of the facings and prolong the life of the clutch.

Look back over the previous section and write out a list of the key bullet points.

Use the interactive media search tools to look for pictures and videos to examine the subject in this section in more detail.

9.1.3 Gearbox

A transmission system **gearbox** is required because the power of an engine consists of speed and torque. Torque is the twisting force of the engine's crankshaft and speed refers to its rate of rotation. The transmission can adjust the proportions of torque and speed that is delivered from the engine to the drive shafts. When torque is increased, speed decreases and when speed is increased, the torque decreases. The transmission also reverses the drive and provides a neutral position when required.

Types of gear Helical gears are used for almost all modern gearboxes. They run more smoothly and are quieter in operation. Earlier 'sliding mesh' gearboxes used straight cut gears, as these were easier to manufacture. Helical gears do produce some sideways force when operating, but this is dealt with by using thrust bearings.

Gearbox For most light vehicles, a gearbox has five forward gears and one

Figure 9.11 Straight cut and helical gears

reverse gear. It is used to allow operation of the vehicle through a suitable range of speeds and torque. A manual gearbox needs a clutch to disconnect the engine crankshaft from the gearbox while changing gears. The driver changes gears by moving a lever, which is connected to the box by a mechanical linkage.

Power, speed and torque The gearbox converts the engine power by a system of gears, providing different ratios between the engine and the wheels. When the vehicle is moving off from rest, the gearbox is placed in first, or low gear. This produces a high torque but low wheel speed. As the car speeds up, the next higher gear is selected. With each higher gear, the output turns faster but with less torque.

Figure 9.12 Pontiac six-speed gear selector

Top gears Fourth gear on most rear-wheel drive light vehicles is called direct drive, because there is no gear reduction in the gearbox. In other words, the gear ratio is 1:1 The output of the gearbox turns at the same speed as the crankshaft. For front-wheel drive vehicles, the ratio can be 1:1 or slightly different. Most modern light vehicles now have a fifth gear. This can be thought of as a kind of overdrive because the output always turns faster than the engine crankshaft.

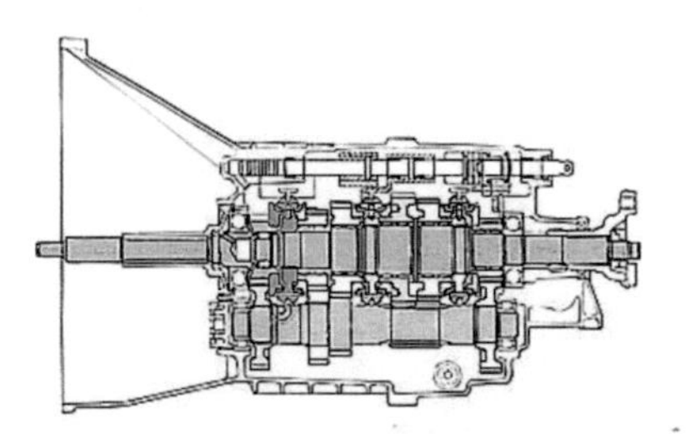

Figure 9.13 Fourth gear is often 'straight through'

Gearbox input Power travels in to the gearbox via the input shaft. A gear at the end of this shaft drives a gear on another shaft called the countershaft or layshaft. A number of gears of various sizes are mounted on the layshaft. These gears drive other gears on a third motion shaft also known as the output shaft.

Older vehicles used sliding-mesh gearboxes. With these gearboxes, the cogs moved in and out of contact with each other. Gear changing was, therefore,

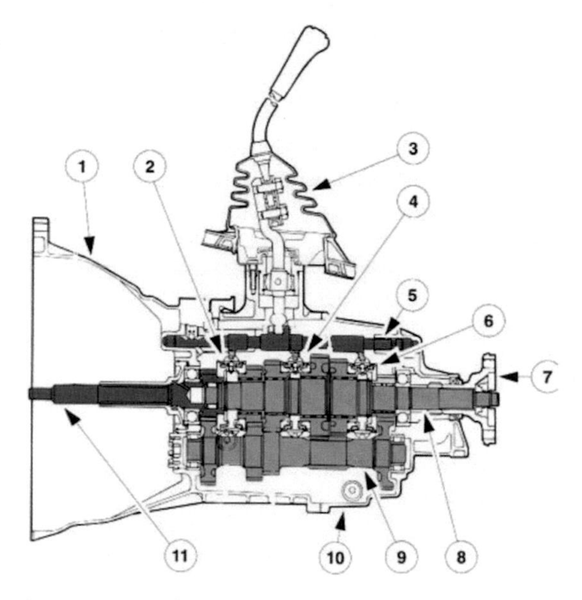

Figure 9.14 Sectioned view of a gearbox

a skill that took time to master! These have now been replaced by constant mesh gearboxes.

Constant mesh The modern gearbox still produces various gear ratios by engaging different combinations of gears. However, the gears are constantly in mesh. For reverse, an extra gear called an idler operates between the countershaft and the output shaft. It turns the output shaft in the opposite direction to the input shaft.

Note how in each case, with the exception of reverse, the gears do not move. This is why this type of gearbox has become known as constant mesh. In other words the gears are running in mesh with each other at all times.

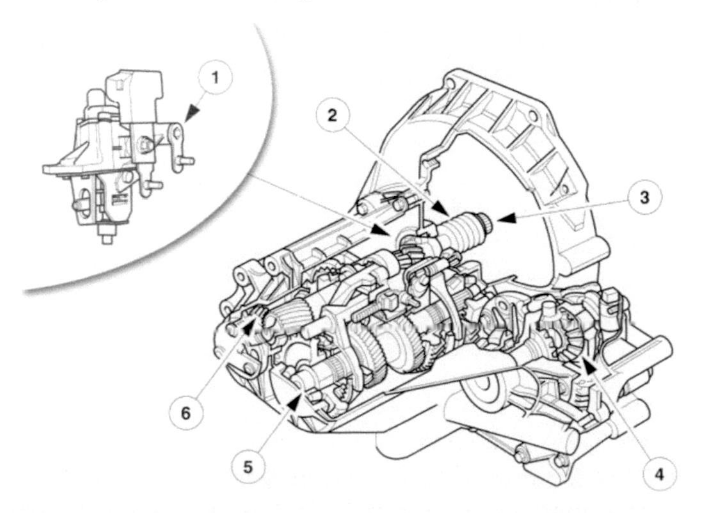

Figure 9.15 FWD gearbox (transaxle)

Remember! There is a multimedia version of this textbook that includes additional images and interactive features: www.automotivett.org

In constant mesh boxes, dog clutches are used to select which gears will be locked to the output shaft. These clutches, which are moved by selector levers, incorporate synchromesh mechanisms.

A manual gearbox allows the driver to select the gear appropriate to the driving conditions. Low gears produce low speed but high torque; high gears produce higher speed but lower torque.

Look back over the previous section and write out a list of the key bullet points.

Create a mind map to illustrate the important features of a component or system in this section.

Introduction There is a wide range of gearboxes in use. However, although the internal components differ, the principles remain the same. The examples in this section are, therefore, useful for learning the way in which any gearbox works.

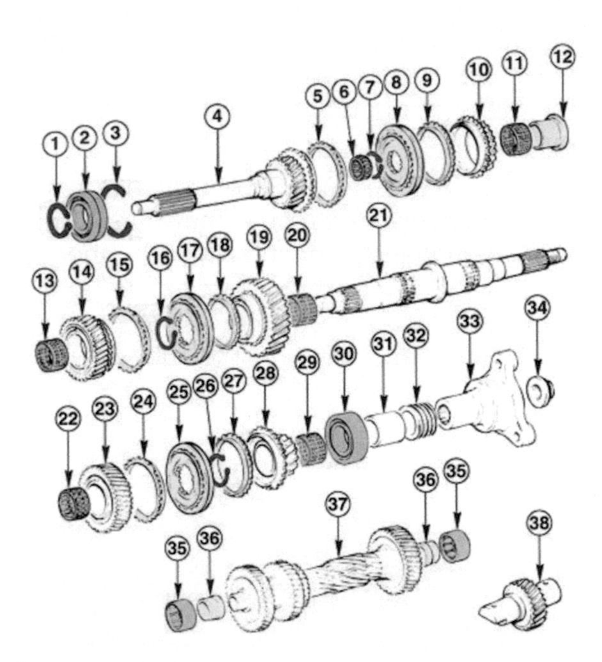

Figure 9.16 Gearbox components

Input shaft The input shaft transmits the torque from the clutch, via the countershaft to the transmission output shaft. It runs inside a bearing at the front and has an internal bearing, which runs on the mainshaft, at the rear. The input shaft carries the countershaft driving gear and the synchroniser teeth and cone for fourth gear.

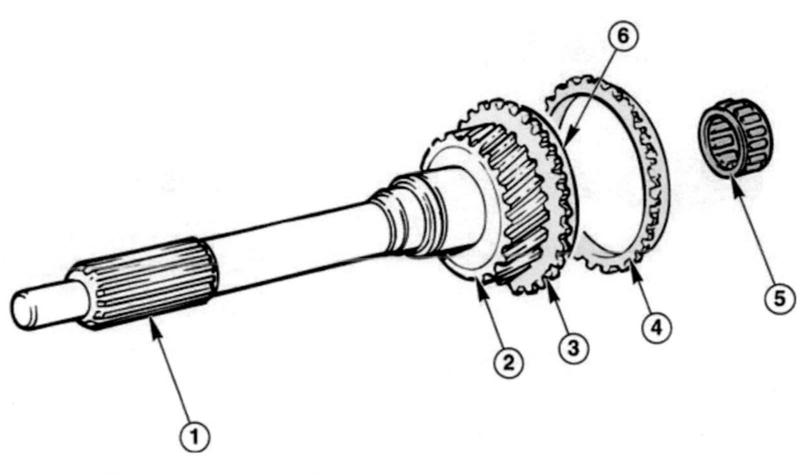

Figure 9.17 Details of the input shaft

Mainshaft or output shaft The mainshaft is mounted in the transmission housing at the rear and the input shaft at the front. This shaft carries all the main forward gears, the selectors and clutches. All the gears run on needle roller bearings. The gears run freely unless selected by one of the synchroniser clutches.

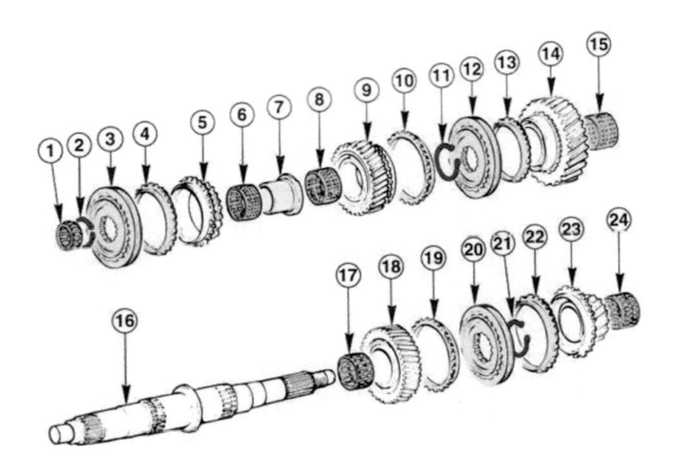

Figure 9.18 Details of the mainshaft

Countershaft The countershaft is sometimes called a layshaft. It is usually a solid shaft containing four or more gears. Drive is passed from here to the output shaft, in all gears except fourth. The countershaft runs in bearings, fitted in the transmission case, at the front and rear.

Reverse idler gear An extra gear has to be engaged to reverse the direction of the drive. A low ratio is used for reverse, even lower than first gear in many cases. The reverse idler connects the reverse gear to the countershaft.

9

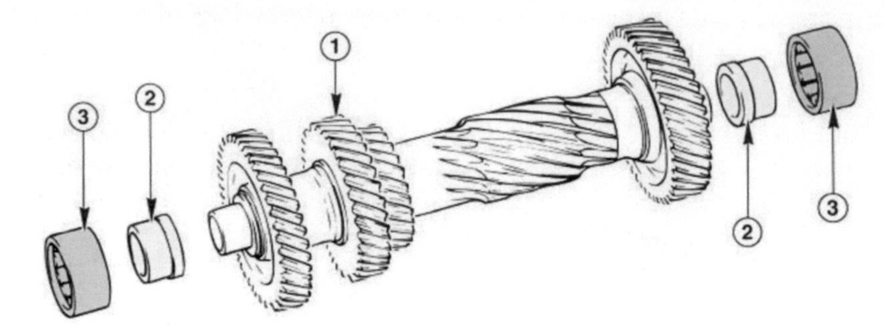

Figure 9.19 Details of the countershaft

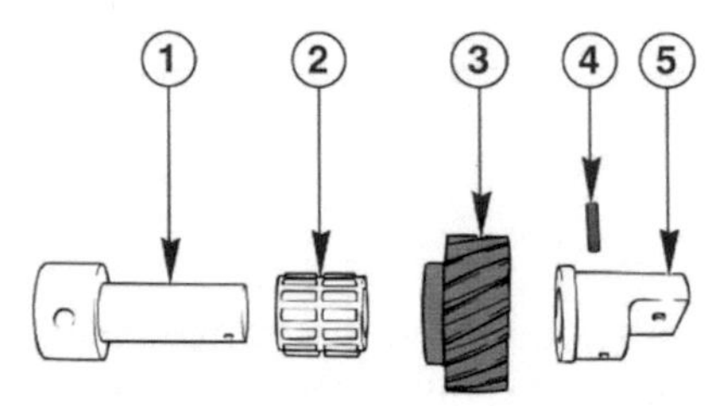

Figure 9.20 Details of the reverse gear idler

Selector mechanism An interlock is used on all gearboxes, to prevent more than one gear being selected at any one time. If this were not prevented, the gearbox would lock, as the gears would be trying to turn the output at two different speeds – at the same time. The selectors are 'U' shaped devices that move the synchronisers.

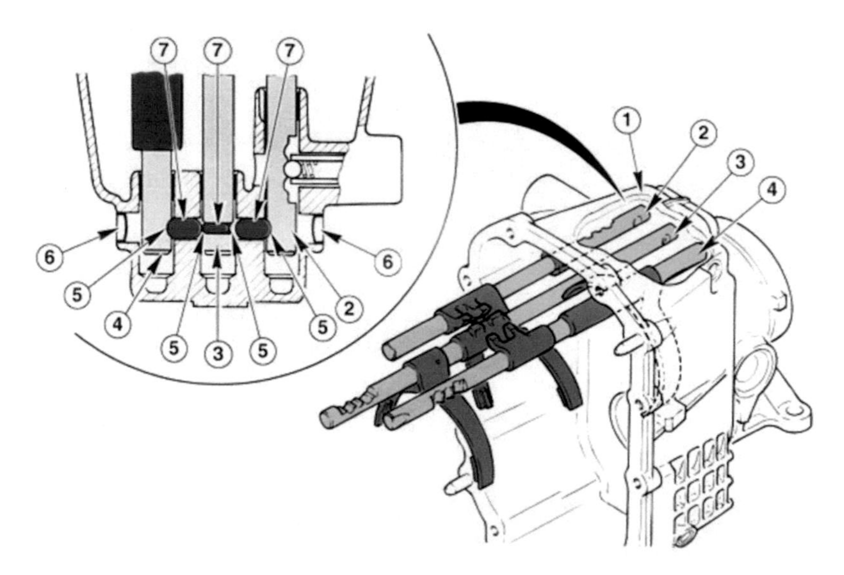

Figure 9.21 Three rail selector mechanism

Transmission fluid The transmission fluid must meet the following requirements:

- Viscosity must be largely unaffected by temperature.
- High ageing resistance (gearboxes are usually filled for life).
- Minimal tendency to foaming.
- Compatibility with different sealing materials.

Only the specified transmission fluid should be used when topping up or filling after dismantling and reassembly. Bearing and tooth flank damage can occur if this is disregarded.

Summary The transmission gearbox on all modern cars is a sophisticated component. However, the principle of operation does not change because it is based on simple gear ratios and clutch operation. Most current gearboxes are five speed, constant mesh and use helical gears.

Answer the following questions either here in your book or electronically.

1 State the purpose of a clutch.
2 State the reason why a gearbox is used.

Look back over the previous section and write out a list of the key bullet points.

Use a library or the interactive web search tools to examine the subject in this section in more detail.

9.1.4 Automatic transmission

Introduction An automatic gearbox contains special devices that automatically provide various gear ratios, as they are needed. Most automatic gearboxes have three or four forward gears and one reverse gear. Instead of a gearstick, the driver moves a lever called a selector. Most automatic gearboxes now have selector positions for Park, Neutral, Reverse, Drive, 2 and 1 (or 3, 2 and 1 in some cases). The fluid flywheel or torque converter is the component that makes automatic operation possible.

Automatic gearbox For ordinary driving, the driver moves the selector to the 'Drive' position. The transmission starts out in the lowest gear and automatically shifts into higher gears as the car picks up speed. The driver can use the lower positions of the gearbox for going up or down steep hills or

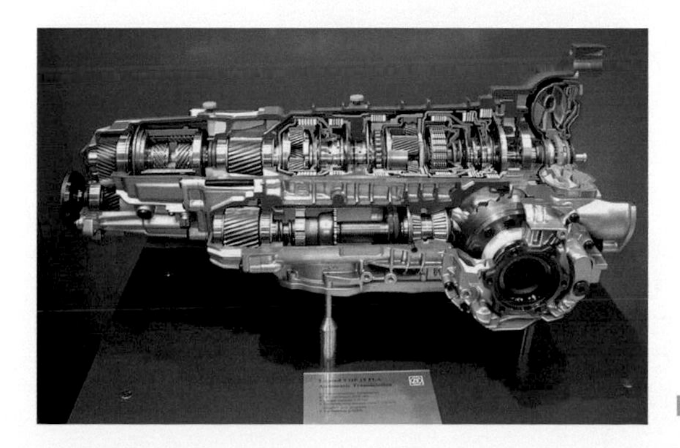

Figure 9.22 Modern auto-box

driving through mud or snow. When in position 3, 2 or 1, the gearbox will not change above the lowest gear specified. Shown here is a modern automatic gearbox used on rear-wheel drive vehicles.

A **fluid flywheel** consists of an impeller and turbine, which are immersed in oil. They transmit drive from the engine to the gearbox. The engine driven impeller faces the turbine, which is connected to the gearbox. Each of the parts, which are bowl-shaped, contains a number of vanes. They are both a little like half of a hollowed out orange facing each other. When the engine is running at idle speed oil is flung from the impeller into the turbine, but not with enough force to turn the turbine. As engine speed increases so does the energy of the oil. This increasing force begins to move the turbine and hence the vehicle. The oil gives up its energy to the turbine and then recirculates into the impeller at the centre starting the cycle over again. As the vehicle accelerates the difference in speed between the impeller and turbine reduces until the slip is about 2%.

Fluid flywheel development A problem, however, with a basic fluid flywheel is that it is slow to react when the vehicle is moving off from rest. This can be improved by fitting a reactor or stator between the impeller and turbine. We now know this device as a torque converter. All modern cars, fitted with automatic transmission, use a torque converter.

Torque converter The torque converter delivers power from the engine to the gearbox like a basic fluid flywheel, but also increases the torque when the car begins to move. Similar to a fluid flywheel, the torque converter resembles a large doughnut sliced in half. One half, called the pump impeller, is bolted to the drive plate or flywheel. The other half, called the turbine, is connected to the gearbox-input shaft. Each half is lined with vanes or blades. The pump and the turbine face each other in a case filled with oil. A bladed wheel called a stator is fitted between them.

Figure 9.23 Torque converter

The engine causes the pump (impeller) to rotate and throw oil against the vanes of the turbine. The force of the oil makes the turbine rotate and send power to the transmission. After striking the turbine vanes, the oil passes through the stator and returns to the pump. When the pump reaches a specific rate of rotation, a reaction between the oil and the stator increases the torque. In a fluid flywheel, oil returning to the impeller tends to slow it down. In a torque converter, the stator or reactor diverts the oil towards the centre of the impeller for extra thrust.

When the engine is running slowly, the oil may not have enough force to rotate the turbine. However, when the driver presses the accelerator pedal, the engine runs faster and so does the impeller. The action of the impeller increases the force of the oil. This force gradually becomes strong enough to rotate the turbine and moves the vehicle. Torque converters can double the applied torque when moving off from rest. As engine speed increases, the torque multiplication tapers off until at cruising speed there is no increase in torque. The reactor or stator then freewheels on its one-way clutch at the same speed as the turbine.

Construct a crossword puzzle using important words from this section. Hint: use the ATT glossary where you can copy the words and definitions (clues!). About 20 words is a good puzzle.

Introduction The main parts of the automatic transmission system are the:

- torque converter with converter lock-up clutch
- fluid pump with stator support
- planetary gear train with clutches and brakes
- intermediate gear stage
- final drive assembly (if FWD)
- control unit or valve body assembly.

Epicyclic gearbox Operation epicyclic gears are a special set of gears that are part of most automatic gearboxes. In their basic form they consist of three main elements:

1. A sun gear, located in the centre.
2. The carrier that holds two, three, or four planet gears, which mesh with the sun gear and revolve around it.
3. An internal gear or annulus, which is a ring with internal teeth; it surrounds the planet gears and meshes with them.

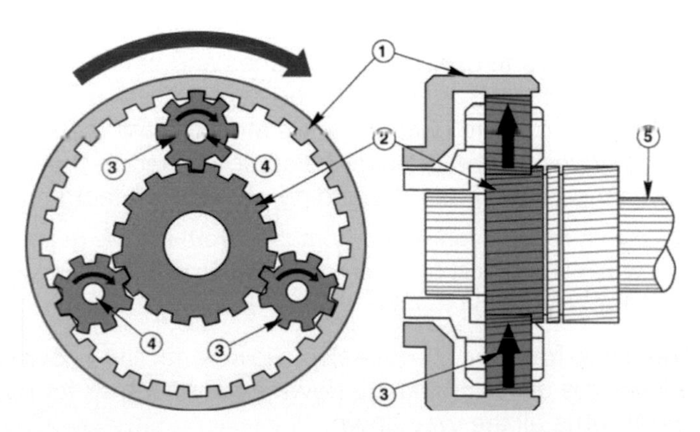

Figure 9.24 Epicyclic gears

Planetary gears Any part of a set of planetary gears can be held stationary or locked to one of the others. This will produce different gear ratios. Most automatic gearboxes have two sets of planetary gears that are arranged in line. This provides the necessary number of gear ratios.

Automatic transmission system As the gear selector is moved into different positions, the power flow through the gearbox changes. All the components shown on this screen are the same as those on the next screen. However, to save space, the labels are not shown so you may need to return here to check details of the power flow.

Power flow The power flows shown here are a representation of what occurs in an auto-box. Note in particular that only the top half is shown! In other words, the complete picture would include a reflection of what is represented here. On this screen, click each button in turn to show how the power flow changes.

Valves and brake bands The appropriate elements in the gear train are held stationary by a system of hydraulically operated brake bands and clutches. These are worked by a series of hydraulically operated valves, usually in the lower part of the gearbox.

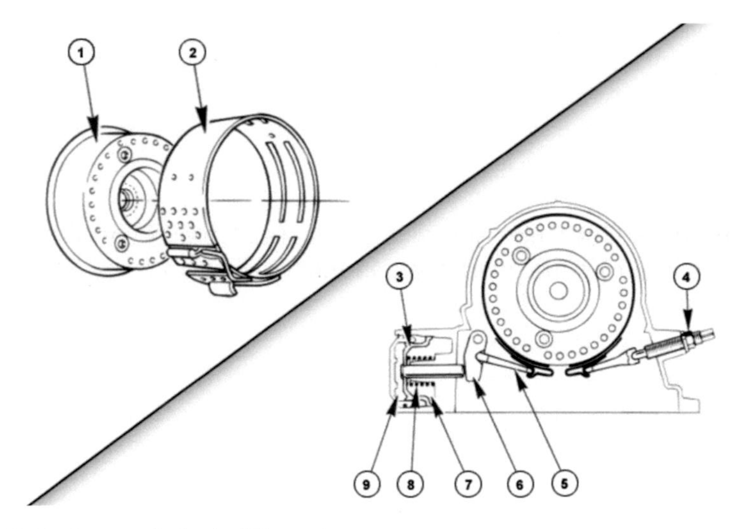

Figure 9.25 Brake bands lock different components

Oil pressure Oil pressure to operate the clutches and brake bands, is supplied by a pump. The supply for this is the oil in the sump of the gearbox. Three forward gears and one reverse gear are achieved from two sets of epicyclic gears. Unless the driver moves the gear selector to operate the valves, automatic gear changes are made depending on just two factors:

1. Throttle opening – a cable is connected from the throttle to the gearbox.
2. Road speed – when the vehicle reaches a set speed a governor allows pump pressure to take over from the throttle.

Kick down The cable from the throttle also allows a facility known as 'kick down'. This allows the driver to change down a gear such as for overtaking, by pressing the throttle all the way down.

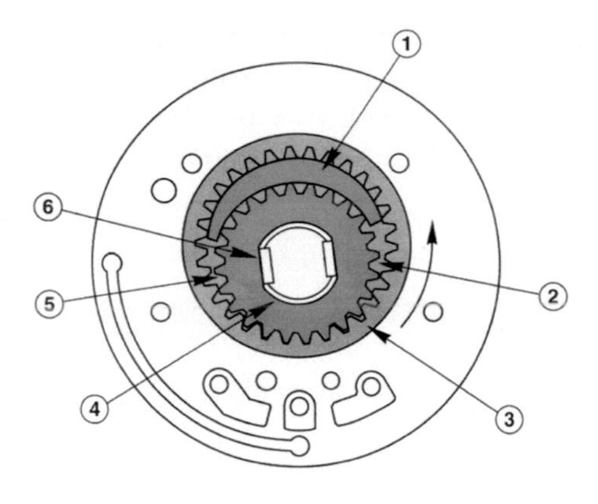

Figure 9.26 Oil pump and governor

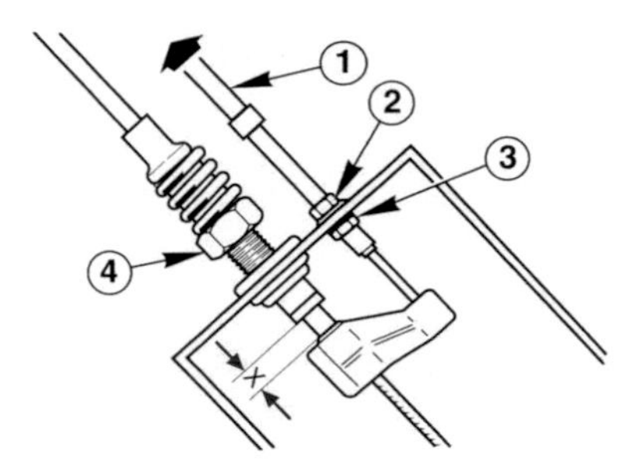

Figure 9.27 Cable position

Standard gear systems Many automatic transaxle gearboxes, use gears the same as in manual boxes. The changing of ratios is similar to the manual operation except that hydraulic clutches and valves are used. Shown here is an example of this system.

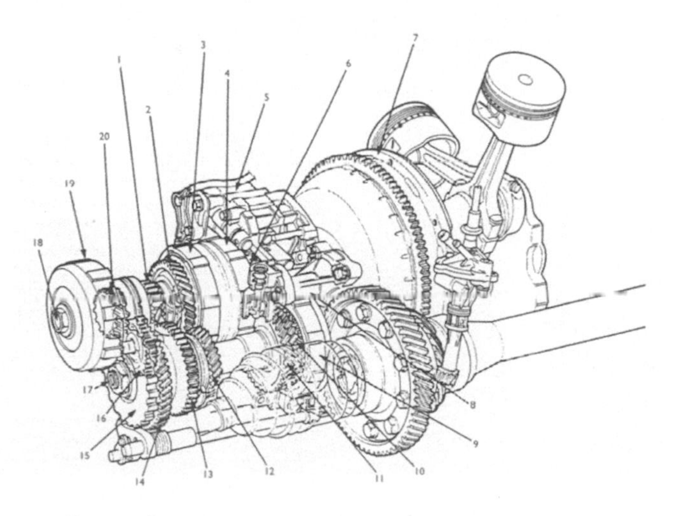

Figure 9.28 Automatic gearbox

Summary Cars fitted with modern automatic transmission systems are a pleasure to drive. Traditionally auto-box cars used more fuel than those with manual transmission. However, the difference is now very small. The main reason is the ability to lock the converter and, therefore, eliminate slip.

Answer the following question either here in your book or electronically.

1 State the purpose AND describe the operation of a brake band.

Look back over the previous section and write out a list of the key bullet points.

Create a mind map to illustrate the important features of a component or system in this section.

9.1.5 Propshafts and driveshafts

Introduction Propshafts, with universal joints, are used on rear or four-wheel drive vehicles. They transmit drive from the gearbox output to the final drive in the rear axle. Drive, then continues through the final drive and differential, via two half shafts to each rear wheel.

Figure 9.29 Propshaft

Main shaft A hollow steel tube is used for the main shaft. This is lightweight, but will still transfer considerable turning forces. It will also resist bending forces.

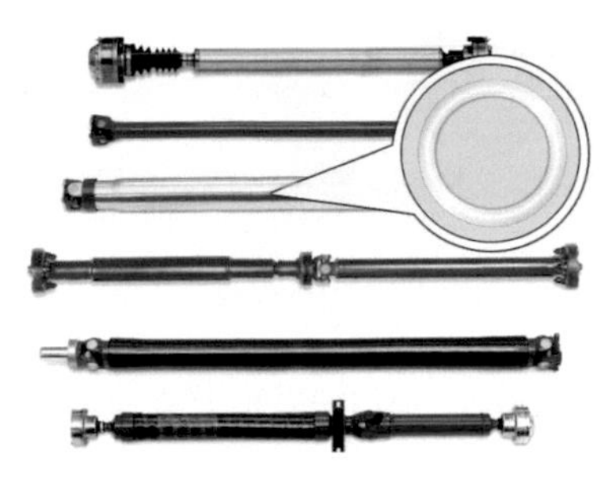

Figure 9.30 Section of a propshaft

Universal joints (UJs) Universal joints, allow for the movement of the rear axle with the suspension, while the gearbox remains fixed. Two joints are used on most systems and must always be aligned correctly.

Figure 9.31 Details of a UJ

Because of the angle through which the drive is turned, a variation in speed results. This is caused because two arms of the UJ rotate in one plane and two in another. The cross of the UJ therefore, has to change position twice on each revolution. However, this problem can be overcome by making sure the two UJs are aligned correctly.

Universal joint alignment If the two UJs on a propshaft are aligned correctly, the variation in speed caused by the first can be cancelled out by the second. However, the angles through which the shaft works must be equal. The main body of the propshaft will run with variable velocity but the output drive will be constant.

UJ bearings The simplest and most common type of UJ consists of a four point cross, which is sometimes called a spider. Four needle roller bearings

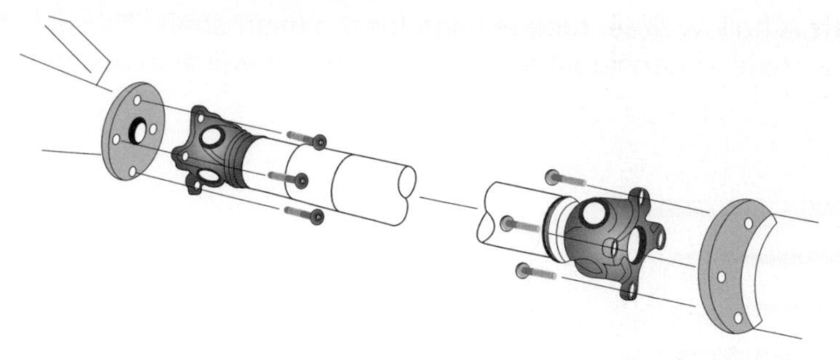

Figure 9.32 These joints are aligned correctly

are fitted, one on each arm of the cross. Two bearings are held in the driver yoke and two in the driven yoke.

Figure 9.33 Details of a universal joint

As the suspension moves up and down, the length of the driveline changes slightly. As the rear wheels hit a bump, the axle moves upwards. This tends to shorten the driveline. The splined sliding joint allows for this movement.

Sliding joint A sliding joint allows for axial movement. However, it will also transfer the rotational drive. Internal splines are used on the propshaft so that the external surface is smooth. This allows an oil seal to be fitted in to the gearbox output casing.

Figure 9.34 A splined joint connects to the gearbox

Centre bearings When long propshafts are used, there is a danger of vibration. This is because the weight of the propshaft can cause it to sag slightly and therefore 'whip' (like a skipping rope) as it rotates. Most centre bearings are standard ball bearings mounted in rubber.

Figure 9.35 This bearing prevents propshaft whip

Summary Propshafts are used on rear or four-wheel drive vehicles. They transmit drive from the gearbox output to the rear axle. Most propshafts contain two universal joints (UJs). A single joint produces rotational velocity variations, but this can be cancelled out if the second joint is aligned correctly. Centre bearings are used to prevent vibration due to propshaft whip.

Construct a word search grid using some key words from this section. About 10 words in a 12 × 12 grid is usually enough.

Driveshafts with constant velocity joints transmit drive from the output of the final drive and differential, to each front wheel. They must also allow for suspension and steering movements.

Constant velocity (CV) joint A CV joint is a universal joint, however, it is constructed so that the output rotational speed is the same as the input speed. The speed rotation remains constant even as the suspension and steering move the joint.

Inner and outer joints The inner and outer joints have to perform different tasks. The inner joint has to plunge in and out, to take up the change in length as the suspension moves. The outer joint has to allow suspension and steering movement up to about 45°. A solid steel shaft transmits the drive.

Figure 9.36 Outer CV joint

Remember! There is a multimedia version of this textbook that includes additional images and interactive features: www.automotivett.org

Figure 9.37 Inner CV joint

CV joint operation When a normal UJ operates, the operating angle of the cross changes. This is what causes the speed variations. A CV joint spider (or cross) operates in one plane because the balls or rollers are free to move in slots. The cross bisects the driving and driven planes.

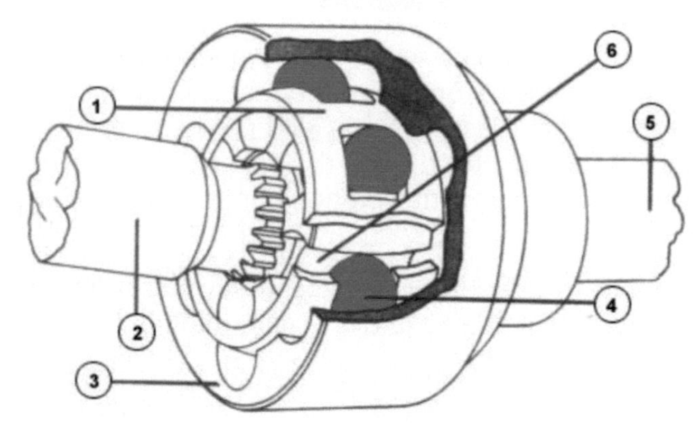

Figure 9.38 Details of a CV joint

The rubber boot or gaiter is to keep out the dirt and water, and keep in the lubricant. Usually a graphite or molybdenum grease is used but check the manufacturer's specifications to be sure. There are a number of different types of constant velocity joint. The most common is the Rzeppa (pronounced reh-ZEP-ah). The inner joint must allow for axial movement due to changes in length as the suspension moves.

Summary Driveshafts with CV joints are used on front-wheel drive vehicles. They transmit drive from the differential, to each front wheel. They must also allow for suspension and steering movements. Inner joints must 'plunge' to allow for changes in length of the shaft. Several types of CV joint are used. All types work on the principle of bisecting the drive angle to produce a constant velocity output.

Look back over the previous section and write out a list of the key bullet points.

Construct a crossword puzzle using important words from this section. Hint: use the ATT glossary where you can copy the words and definitions (clues!). About 20 words is a good puzzle.

9.1.6 Final drive

Introduction Because of the speed at which an engine runs, and in order to produce enough torque at the road wheels, a fixed gear reduction is required. This is known as the final drive. It consists of just two gears. The final drive is fitted after the output of the gearbox on front-wheel drive vehicles. It is fitted in the rear axle after the propshaft on rear-wheel drive vehicles. The ratio is normally between about 2:1 and about 4:1. In other words, at 4:1, when the gearbox output is turning at 4000 rev/min, the wheels will turn at 1000 rev/min.

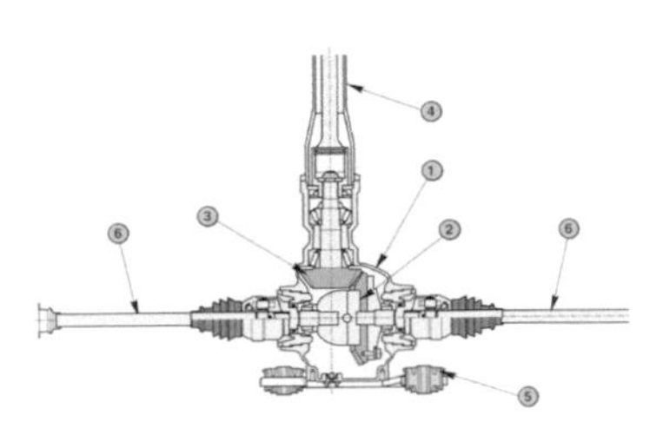

Figure 9.39 Rear axle final drive gears

Rear-wheel drive The final drive gears turn the drive through 90° on rear-wheel drive vehicles. Four-wheel drive vehicles will also have this arrangement as part of the rear axle.

Front-wheel drive Most cars now have a transverse engine, which drives the front wheels. The power of the engine therefore does not have to be carried through a right angle to the drive wheels. The final drive contains ordinary reducing gears rather than bevel gears.

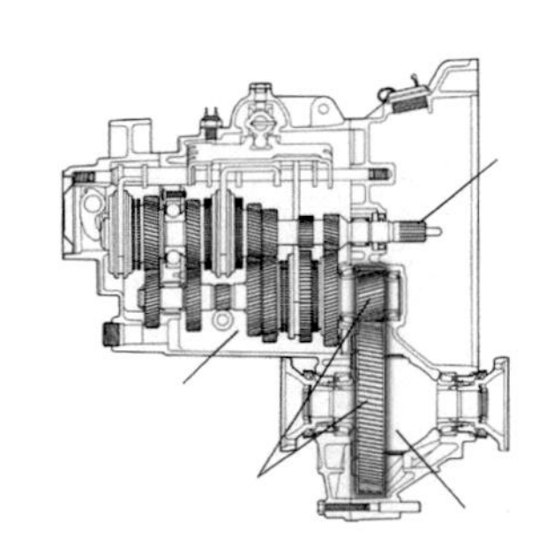

Figure 9.40 Transaxle final drive gears

Reduced speed and increased torque Final drive gears reduce the speed from the propeller shaft and increase the torque. The reduction in the final drive multiplies any reduction that has already taken place in the transmission.

Hypoid gear The crown wheel gear, of a rear-wheel drive system, is usually a hypoid type, which is named after the way the teeth are cut. As well as quiet operation, this allows the pinion to be set lower than the crown wheel centre, thus saving space in the vehicle because a smaller transmission tunnel can be used.

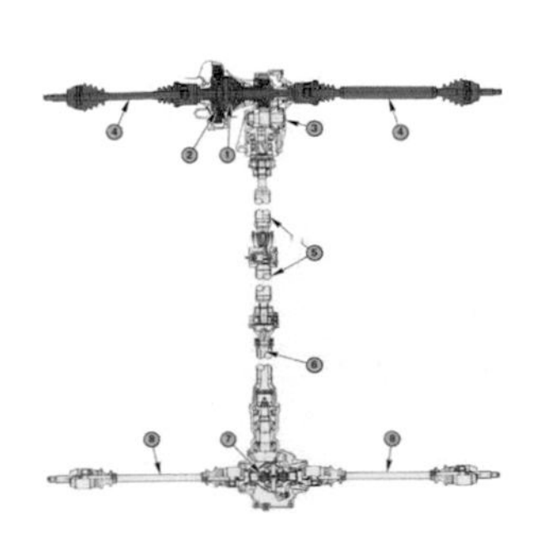

Figure 9.41 4WD final drive layout

Hypoid gear oil Because the teeth of hypoid gears cause 'extreme pressure' on the lubrication oil, a special type is used. This oil may be described as 'Hypoid Gear Oil' or 'EP', which stands for extreme pressure. As usual, refer to manufacturers' recommendations when topping up or changing oil.

The complete rear axle assembly consists of other components as well as the final drive gears. The other main components are the differential, halfshafts and bearings. Components that make up a solid axle are shown here. Some rear-wheel drive and four-wheel drive vehicles have a split axle. On these types, the final drive is mounted to the chassis and driveshafts are used to connect to the wheels.

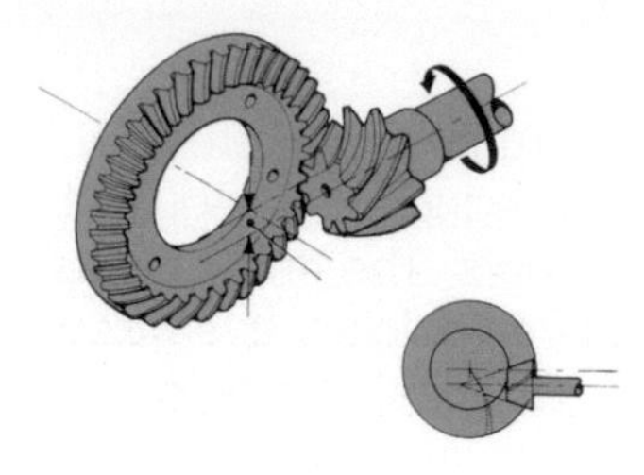

Figure 9.42 The design allows a lower propshaft to be used

The front-wheel drive axle, where a transaxle system is used, always consists of the final drive and two driveshafts. The gearbox, final drive and one driveshaft are shown here. The final drive gears provide the same reduction, as those used on rear-wheel drives, but do not need to turn the drive through 90°.

Figure 9.43 Lubrication is important

Four-wheel drive The general layout of a four-wheel drive system is shown here. A representation of how torque is distributed is also shown. The variation in torque is achieved by differential action. This is examined in some detail later in this programme.

Summary To produce enough torque at the road wheels, a fixed gear reduction is required. This is known as the final drive. It consists of just two gears. On rear-wheel drive systems, the gears are bevelled to turn the drive through 90°. On front-wheel drive systems, this is not necessary. The drive ratio is similar for front or rear-wheel drive cars.

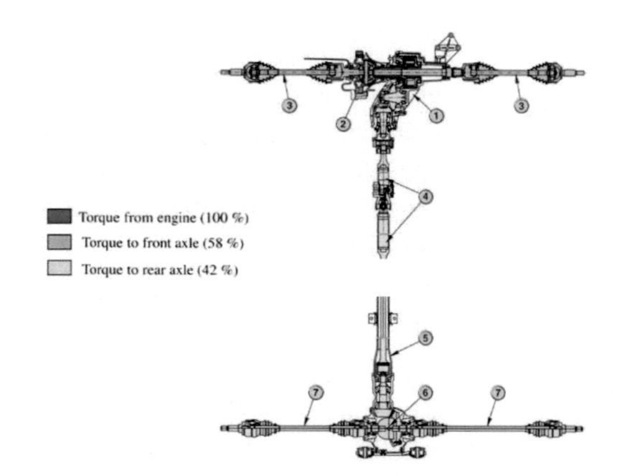

Figure 9.44 Torque distribution in a 4WD system

Look back over the previous section and write out a list of the key bullet points.

Create an information wall to illustrate the important features of a component or system in this section.

9.2 Transmission systems puzzles

9.2.1 Cryptograms

A	B	C	D	E	F	G	H	I	J	K	L	M	N	O	P	Q	R	S	T	U	V	W	X	Y	Z
		9											6												

	C				C													C		N	N		C			N	
19	9	2	8	11	9	10		26	3		25	7	18		15	26	3	9	7	6	6	12	9	11	26	6	16

	N			C		N	N		C			N								N								
19	9	15		9	7	6	6	12	9	11	26	6	16		18	7	11	19	26	6	16		3	10	19	25	11	9

Cryptogram 9.1 Take up the drive:

9.2.2 Anagrams

Martinis Sons
Stains Minors
Mansion Stirs
Roman Insists

Anagrams 9.1 On the radio: what word did I use to get these anagrams?

9.2.3 Word search

C E G B Z K X S L I R J D Z T
C R L O R C X E M L E M Q V Y
N Y E B Q U E Q G I V B T C G
S E L C A H K S O L E N O I D
S Q L I W C O Z H N L R C N V
L G M Y N O X L L C B T K O W
B T L G I D K S H Q B F Q R D
Z F L E I I E H I Q V C O T J
B G P A S H F R D C L Z L C K
M Q O R H K K Q C U D T F E S
Q V S B L A N O T S I P H L V
T E N O I Z R C V C A Z B E H
T Q M X C O H D T E S Q K Y R
G F R C I G D I C U F D Q S D
P Y B M J I Y X D D M V H M F

CABLE
CLUTCH
CYLINDER
ELECTRONIC
FLYWHEEL
GEARBOX
LEVER
PISTON
SOLENOID

Word search 9.1 Clutch

9.2.4 Crossword

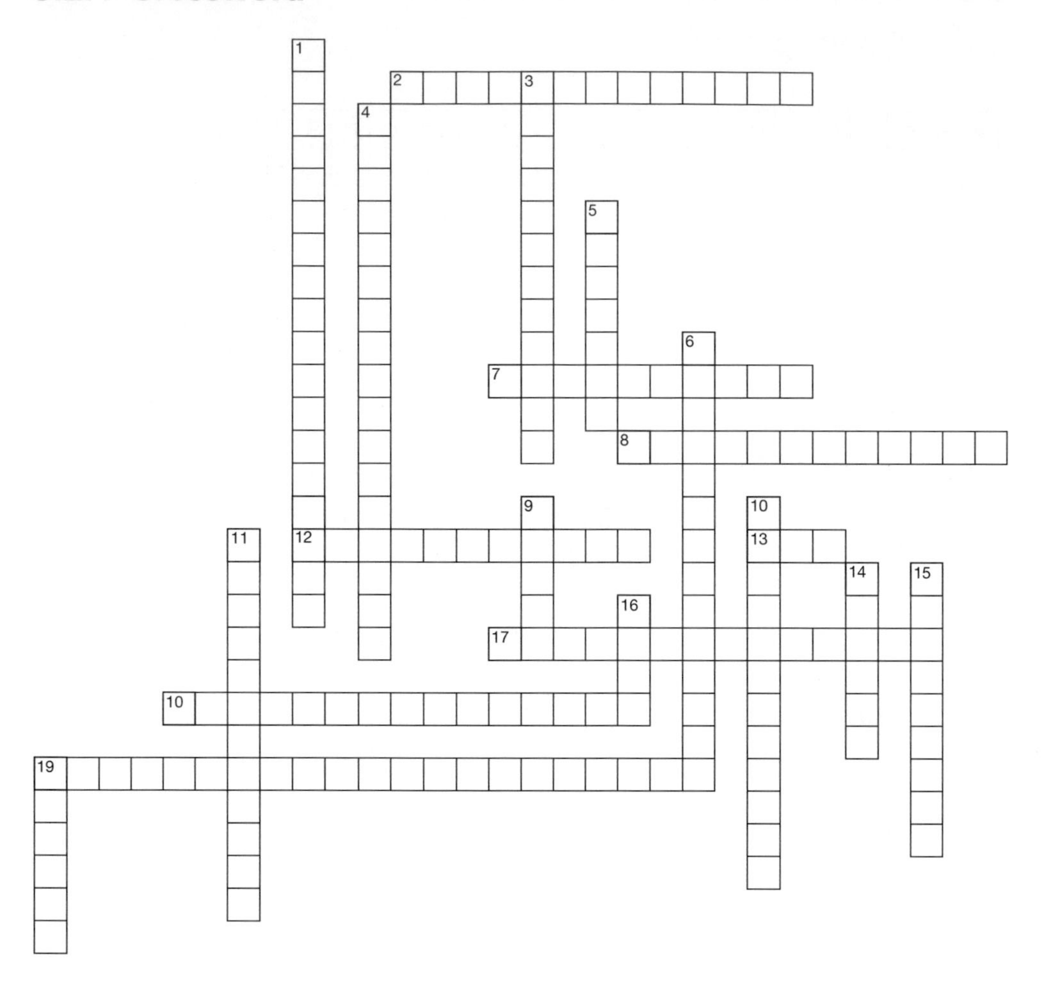

Across

2 A drive coupling that uses a liquid to transmit drive (5,9)

7 Fixed gear reduction in transmission and can turn rotation through a right angle (5,5)

8 A series of conical rollers held in a cage between inner and outer tracks (5,7)

12 Inner and outer tracks with spheres in between to allow smooth rotation (4,7)

13 Centre of a wheel or gear or similar

17 Rotating shaft coupling that permits a change in angle (9,5)

Down

1 DSG (6,5,7)

3 Gear set that allows two wheels to rotate at different speeds

4 The oil used in a automatic gearbox (12,5)

5 Gearbox position when no gear is selected and no drive transmitted

6 Connects gearbox to final drive and transmits torque on rear-wheel drive (10,5)

9 wheel Bevel gear in which the teeth are set around the outside, used in a final drive (5,5)

10 Roller or balls used to support a drive shaft or component on an axle (5,7)

Across

18 A fluid flywheel with a stator that increases turning force (6,9)

19 A coupling that transmits drive through an angle without changes in speed (8,8,5)

Down

11 Part of gear change system that makes gear speeds the same before they are engaged

14 A rubber or similar boot used to cover working components

15 Gearbox mechanism that prevents the engagement of two gears at the same time

16 Toothed wheel that meshes with another toothed wheel

19 A device to allow drive to be connected and disconnected

Crossword 9.1 Clutch, manual gearbox, automatic transmission, driveline, final drive and differential

Index

abrasives compounds 106–7
accident procedure 28
accident repair reception 35
accidents, inspection after 98, 99
accuracy 57–9
ACEA (Constructeurs Européens d'Automobiles) 138, 139
Ackermann, Rudolf 212–13
Ackermann steering 212–13
adhesives 78
adjustable spanner (wrench) 51
air chisel 53
air conditioning (AC) servicing unit 66, 68
air cooled system 130, 131
air cooling 133
air cutter 53
air filters 144–5
air flow (into diesel engine) 156
air flow meter 150
air fresheners 110
air gun 65
air supply components 144
air temperature sensor 153
air wrench 51
Allen head screw 77, 80
alloy wheels 235–6
alternator 87, 88, 125
American Petroleum Institute (API) viscosity ratings 138, 139
angle locator 62
anti-roll bar 203
antifreeze 133
antilock brake system (ABS) 225
API (American Petroleum Institute) viscosity ratings 138, 139
aspect ratio 24
automatic gearbox 259–61
automatic static actuators 188
automatic transmission systems 259–64
automobile industry 31–47
auxiliary air valve 145
Avon cost conscious tyre 243, 244
axle stand 65, 66
axles 89

BA (British Association) system 78
baffle silencer 163
balancer 71
ball joint 216, 217
ball joint presses 67–8
ball joint splitter 68
battery 125–6
battery chemistry 126
beam axle 202
beam setter 188
bearing puller 54
bimetal strip 179
blade (engineer's) screwdriver 51
body components 46
body design 45
bodyshop 34
bodywork protection 97
bolt heads 76
bolts 73, 75, 76
Bosch VR pumps 160, 161
  high-voltage magneto ignition system 168
  injection pump 158
bottom dead centre (BDC) 118
boundary lubrication 137
brake adjusting tools 54–5
brake adjustment hole 230
brake adjustments 228–9
brake lights 190
brake pads 224
brake pipes 226, 227
brake servo 225
brake shoes 224, 225
brakes
  requirements 99
  servicing 96
braking system checks, routine 222–32
  test requirements 227
  brake adjustments 228–9
  manual adjustment 229–31
breaker points 169
Bristol screw head 77
British Association (BA) system 78
British Standard Whtworth 78
bulbs 184–5
bump stop 202
button screw head 77
bypass air duct 151
bypass system 130
bypass-mixing cooling system 134

cable size 180
cables 179
  UK colour code 180
camshaft 118
canister purge solenoid valve 149
capacitor 169
captive spring nut 79, 80
carpet, cleaning 110
cast alloy wheels 236
castellated nut 75
catalytic converter 157, 158
cavitation 135
CCMC (Comité des Constructeurs d'Automobile du Marché Commun) 139
central point (CPI) systems 143
centre bearings 266
centre bolt 207
centre high mounted stop lamps (CHMSL) 192
centrifugal advance 169
chain cutter 53
chain wrench 53
charging system 125–6
chassis 86, 87–9
  integrated 45, 46
  ladder 46
  separate 45
  systems 197–244
circuit breaker 178
circuit diagram 181–2
circuit faults 182–3
circuit protection 177
circular valve timing diagram 118
clay kit 106
clay products 106
clutch 89, 249–53
  basic functions 250
  engagement 251
  frictional facings 252–3
  health hazards 253
  main parts 250
clutch aligner kit 71, 73
clutch cable 249
clutch disc 250, 252
clutch location 250
clutch shaft 252
coil spring compressors 67
coil springs 201, 204, 205, 251
cold start devices 156
collant holes 130
combustion chambers 115
combustion/power stroke 116, 117
Comité des Constructeurs d'Automobile du Marché Commun (CCMC) 139
companies
  franchised dealer, role of 37
  parts department 39
  parts stock 39
  reception and booking systems 37–9
  security 39–40
  structure 37
  types 35, 36–7
compression ignition fuel systems 154–64
  diesel injection 155–62
  exhaust systems 162–4
compression stroke 116, 117

concave reflector 186
concept car 45
condenser 169
connector block 179
connectors 178, 179
conrod big end lubrication 137
constant mesh gearboxes 255–6
constant velocity (CV) joint 267
  operation 268
contact breaker system 168
contact breakers 169
convergent beam 186
convertible car 45
convex reflector 187
coolant 132, 133
coolant circuit 132
coolant density 135
coolant flow 130, 134
coolant port 131
coolant pump 135
coolant thermistor 153
cooling 127–37
cooling components 133–6
cooling fan 133
cooling-system design 132
copper 179
core plus 130
COSHH regulations 102
countershaft 254, 257, 258
countersunk screw head 77
coupe car 45
crank sensor 152
crankcase 121
crankcase ventilation 144–5
crankshaft 86, 118
cross-head screwdrivers 52
current clamp 62
customer contracts 97
cylinder block 121
cylinder charge 118
cylinder components 120–2
cylinder head 130
cylinder liner 120

damper operation 212
dampers 200, 209–11
de-waxing 109
dead blow hammer 55
DI injector 160, 161
dial calipers 61
dial gauge 60–1
diaphragm clutch 250, 252
diesel 114
diesel engines 115–16, 118
diesel fuel injection components 158
diesel fuel injection systems 155–62
digital calipers 61
digital multimeter (DMM) 62
dip beam 189
direct drive 254
direct injection 156, 157
disc brakes 88, 224, 227, 228, 231
distance-based services 97
distributor 169, 170
divergent beams 186
dome (button) screw head 77
dome nut 75
double hex screw head 77
double overhead camshaft engine (DOHC) 124
driven plate 252
driveshaft 89, 264, 267
drum brakes 224, 227, 228, 229
dry liners 120, 121
dry vacuum cleaner 104
dust chamber 144

electric drill 65
electric welder 65
electrical components and circuits 177–83
electrical harness 149
electrical system 86, 175–90
electrics 125–7
electronic control (diesel fuel injection) 157
electronic control unit (ECU) 142, 143, 144
electronic fuel injection (EFI) 142
  components 144
electronic systems 87
emission control 129, 149
emissions 129
employment opportunities 33
energy conversion 224
engine cleaning 108
engine components and operation 113–14
engine control module (ECM) 143–4
engine coolant temperature sensor 152–3
engine crane 17, 65
engine cylinder honing tool 70
engine degreaser 108
engine load 150
engine locations 119
engine mounting in vehicle frame 119
engine operating cycles 114–19
engine speed and position 149
engine system 86, 113–70
engineer's screwdriver 51
epicyclic gearbox 261
equipment maintenance 18–20
estate car 45
exhaust gas oxygen sensor 153, 154
exhaust gas recirculation (EGR) 157
exhaust gas recirculation (EGR) valve 149
exhaust manifold 162
exhaust mountings 163, 164
exhaust removal tools 53–4
exhaust stroke 116, 117
exhaust system 86, 162–4
  components 162
  down pipe or front pipe 163
  exhaust movement 163
  high temperatures 162
  joints 163, 164
  regulations 99
  requirements 162
expansion 132, 135
expansion tank 136
extension nut 75
exterior cleaning 105–9
  abrasives compounds 106–7
  clay products 106
  drying 106
  engine cleaning 108
  engine degreaser 108
  finishing 107
  glass cleaning 107–8
  inspection 106
  new car de-waxing 109
  paint washing 105
  polymers and waxes 107
  rubber and plastic 108–9
  tar and road film removers 106
  washing process 105
  wheels 105
external star washer 79
external thread 76

fast fit centre 35
fasteners 73–80
feeler gauges 59
final drive 269–70
final drive assembly 89
finishing 107
fire 24–6
fire extinguishers 25–6
fire procedure 24–5
fire triangle 24
fire types 25
first service 96
flange connections 164
flange nut 75
flap type air flow meter 150
flat (countersunk) screw head 77
flat washer 79
floor mats 97
fluid flywheel 260
flywheel 86, 89
focal length 186
focal point 186
focused beam 186
Ford steering system 217
four-stroke cycle 116–20
four-wheel drive 43, 270
franchised dealer, role of 37
friction, lubrication and 137–9
front axle 216
  geometry 215
front engine forward-wheel drive (FWD) 42
front engine rear-wheel drive (RWD) 43
front-wheel drive 269–70
front-wheel-drive gearbox (transaxle) 255
fuel filter 146, 147, 159
fuel gauge 141
fuel heating 159
fuel injection methods 142
fuel injection pump 141
fuel injectors 160
fuel lift pump 159–60
fuel pressure regulation 148
fuel pressure regulator 146
fuel priming pump 160
fuel pump 141, 146
  electrical supply 147
fuel rail 146
fuel supply 141, 145–6
  components 144
fuel tank 146

fuses 177–8
fusible links 178

gas damper 209
gas spring 201
gas suspension unit 203
gas welder 65
gas/fluid suspension 202
gasoline direct injection (GDi) 143
gearbox 89, 253–9
  automatic 263
  components 256
  input 254–5
  power, speed and torque 254
gearbox input shaft 252
gears, types of 253–4
General Motors lights 189
glass cartridge, ceramic and blade type. fuses 177
glass cleaning 107–8
glaze buster 70

hammer 51
  dead blow 55
  soft 54
hand tools 19, 51–7
handbrake 88, 231
handbrake linkage 231
hard wax 109
hatchback car 45
hazards 21–2
headlight alignment 187
headlight bulb 185
headlight lenses 186–7
headlight reflectors 185–6
heat distribution 134, 135
heated fuel filter housing 160
helical gears 253
helical springs 204
hex screw head 77
hex socket head screw 77, 80
hexagon socket spanner 51
high intensity discharge lamps (HIDs) 189, 190
high pressure components 158–9
high pressure pump 155
high resistance circuit 183
Highway Code 99
hoist 65
honing tool 70
horn 99
hose clips 52–3
hose removal 52
hot wire/film air flow meter 151
hub 252
hydraulic clutch 249, 250
hydraulic components 125
hydraulic cylinders 224
hydraulic dampers 209
hydraulic pressure 87
hydraulic valve adjustment 124
hypoid gear 269
hypoid gear oil 269

idle air control valve (IAC) 145
idle speed 145
idle speed control valve (ISC) 145
ignition coil 167,l 169
ignition switch 169
ignition systems 164–70
  conventional 166
  developments 167
  early 168
  electronic 166
  energy storage 168
  engine load 167–8
  engine management 166
  high voltage generation 167
  mechanical switching 168
  modern systems 168
  purpose 166
  spark advance 168
  types 166
ignition timing 167
in-line paper element type 147
incorrect adjustments 91–2
independent front and read suspension (IFS/IRS) 202
independent suspension systems 204
indirect injection (IDI) 156, 157, 160, 161
induction/intake stroke 116, 117, 118
inductive speed sensor 150
inertia switch 147
information sources 92–3
infrared lights 189
injection pressures 158
injection pump 158, 160
injection time, calculation of 149
injection wiring harness 149
injector operation 161
injector pulse width 148
injector spray 162
injector valves 148
inner and outer joints 267
input shaft 256, 257
inspection 97, 106
  aural 98
  functional 98
  reasons for 98
  types 98–9
  visual 98
interior cleaning 109–10
internal combustion engine 114, 115
  air and fuel 115
internal star lock washers 78
internal thread 76

jack 65, 66, 67
job cards 93
jumper wire 56

kick down 262

ladder chassis 46
lambda sensor 153
laminated or multi-leaf springs 206
large goods vehicle (LGV) 44
layout circuit diagram 181–2
layshaft 257
lead-acid batteries 126
leads 170–2
leaf spring 201, 206
leaf spring fixings 207
leather, cleaning 109–10
lenses 187
levelling actuator 188
lever-type splitter 68, 69
levers 52, 55
lifting heavy loads 16–17
light emitting diodes 188–9, 192
light van 46
light vehicle construction 41–4
light vehicles (LV) 44
lighting clusters 184
lighting system maintenance 175–92
lights 99
link 202
liquid cooling 133
liquid petroleum gas (LPG) 114
load index (tyre) 243
locking nuts 75–6
Loctite Threadlocker 78, 79
low pressure components 159
low pressure fuel lines 159
lubrication 137–8
lubrication system checks 127, 128, 137–9
  environmental regulations 137–8
Lucas DP pump 160

magneto 168
main dealer 33, 35
main systems 86–9
mainshaft 257, 265
maintenance 85–109
  equipment 18–20
  lighting system 175–92
malfunction indicator light 153
manifold absolute pressure (MAP) 150, 168
manifold absolute pressure (MAP) sensor 151, 152
manoeuvrability 215–16
manufacturers' warranty systems 93
mass air flow meter 151
master cylinder 249
McPherson strut 68, 201
measurement 57–64
mechanical lift pump 141
micrometer 58, 59, 60
  thimble reading 59, 60
microprocessors 125
mid-engine rear- wheel drive 43
monocoque 45
MOT test 97
motorist discount store 35
moving heavy loads 16–17
mufflers 163
multi-leaf springs 206
multimeter 62–3
Mustang engine 93

negative temperature coefficient (NTC) thermistor 152
night vision 189
number plates 99
nuts 73, 74, 75–6, 77, 78

oil pressure 262
oils, engine 138–9
  multigrade 138
  recommended grades 139
  synthetic and semi-synthetic 138
  two stroke 139

one-way screw head 77
open cell battery 126
open circuit 182, 183
open-ended spanner 51
oscilloscope 63
Otto cycle 116–20
output shaft 254, 257
oval (raised head and countersunk) screw head 77
overdrive 254
overhead cam (OHC) layout 123
overhead camshaft (OHC) valve operation 123–4
overhead valve (OHV) 123
overhead valve (OHV) layouts 122
overhead valve (OHV) valve gear 122
overheating 132
oxyacetylene welding equipment 53

paint washing 105
pan screw head 77
Panhard rod 203
paper clip 57
particulates 156
parts department 39
parts stock 39
parts washer 65
passenger carrying vehicle (PCV) 44
pentalobular screw head 77
performance tyre tread 243, 244
personal protective equipment (PPE) 15–16, 102
petrol 114
petrol engines 115
pistons in 115–16
petrol injection 142–54
Phillips head screw 77, 80
Philips screwdrivers 52
pickup truck 46
PicoScope 63–4
pilot/spigot bearing puller 72, 73
pin-in-hex-socket 77
pintle ('pintaux') fuel injectors 160
pipe clamp 69
piston 86, 115–16
pitch 78
planetary gears 262
plastic, cleaning 108–9
plenum chamber 144
pliers 52
plug leads 169
pneumatic tyres 240–1
pole 186
polisher 104
polydrive screw head 77
polymers 107
poppet valves 116
post-work 98
power flow (automatic transmission system) 262
power steering pump 213
power steering systems 214
power train control module (PCM) 143
Pozidrive screwdrivers 52
Pozidrive screw head 77
pre-delivery inspection (PDI) 98, 99
pre-work 98
preload 76
pressed steel wheels 234
pressure bleeder 69–70
pressure cap 136
pressure-cap vacuum valve 136
pressure differential 158
principal axis 186
propshaft 43, 264, 265–7
pump impeller 260
pump types 160

rack and pinion steering 88, 213, 218
radial tyre carcass 241
radial tyre features 241
radiator 133
ramp 65
rear axle 269–70
rear engine rear-wheel drive 43–4
rear lights 188
rear-wheel drive 43–4, 254, 260, 269–70
reception and booking systems 37–9
records 97
regulations 99, 137–8, 146, 148
relays 177
replaceable liners 121
resistive air flow meter (MAF) 150
reverse idler gear 257, 258
reverse lights 190–2, 191
ring spanners (wrenches) 51
road film removers 106
Road Traffic Act 99
rocker arm systems 124
roller cell pump 147
round screw head 77
routine vehicle checks 85–6
rubber, cleaning 108–9

SAE (Society of Automotive Engineers) viscosity ratings 138, 139
safety 13–28, 102
safety data sheet 23
safety procedures 28
saloon car 45
screw and nut steering box 218
screw heads 76, 77
screw threads 76
screw wedge-type adjusters 124
screwdrivers 76
screws 76, 77, 79
dimensions 78
tightening 78
seat belts 100
seat covers 97
seats 89
secondary air solenoid valve 149
security 39–40
security hex socket (pin-in-hex-socket) 77
selector 259
selector mechanism (gearbox) 258
self-servo action 224
self-tapping 76
self-tapping screws 79
sensors 149
sensors 87
Service and on board diagnostic (OBD) plugs 153
service history 96
service life 132
service sheets 90–2
servicing 95–7
reasons for 96
servicing data 96
servicing requirement books 96
shackles 207
shank 78
shims 124
shock absorbers 200–1
short circuit 182, 183
showroom 33
Sidelights, taillights, brake lights 187
signage 26–7
single leaf or mono-leaf springs 206
single leaf spring 207
single point (SPI) systems 142
single tube telescopic damper 209, 210
slave cylinder 249
slide hammer 72
sliding joint 266
sliding-mesh gearboxes 254–6
slot (flat) screw head 77
slot and hex head self- tapping screw 80
snake-eye screw head 77
Society of Automotive Engineers (SAE) viscosity ratings 138, 139
socket wrench 51
soft hammers 54
soft wax 109
soldering iron 56
solenoid injectors 149
solenoids 145
spanner head screw head 77
spark ignition fuel systems 139
spark plug 169, 170–2
electrode gaps 171
heat range 170–1
high tension (HT) 171–2
HT cables 172
standard 170
temperature 170
special inspection 98, 99
special purpose wrenches 52
speed ratings (tyre) 243
speed-sensitive power steering systems 214
speedometer 100
spider 265
spline drive screw head 77
splined joint 266
split washer 79
spoked wheels 234
sports utility vehicle (SUV) 46
spring (split) washer 79
springs 89, 200, 204–8
square screw head 77
standard tyre tread 243, 244
starter motor 87, 125
starter system 126

steam cleaner 65
steel rule (ruler) 57, 58
steel wheel rim features 234, 235
steering 100, 212–19
components 214–16
construction 216
development 212
necessity for 214
steering alignment 219–22
measure and adjust tracking 219
steering geometry 220
taking readings 220
track rod adjustment 219
tracking (toe in/out) 219
steering gear 216
steering rack 213, 218
steering wheel 88
steering wheel covers 97
stepper motors 145
stoplights 190–2
stores and parts records 93
straight cut gears 253
straight edge and feelers 59–60
strap wrenches 53
strut 201
stub axles 216
stud 76
suspension system 88, 199–204
components 199
compromise in 199
effect of 200
steering and 214–15
suspension bushes 212
swirl 158
swirl chambers 158
switch 177, 190–2
symbols 181
systems 36

tape measure 57, 58
tar remover 106
telescopic damper 201
terminal circuit 183
terminal circuit diagram 182
terminal kit 56
terminals 178, 179
test lamp 55, 56
thermal-imaging systems 189
thermostat 132, 133–4
thread 76
throttle body 144, 145
throttle body injection (TBI) 142
throttle idle speed control 146
throttle plate assembly 144
throttle potentiometer 152
tie-bar 202
tie-rod 202
time-based services 97
timescales, working to 99
tools and equipment 49–73
top dead centre (TDC) 118
top gears 254
topping up, battery 126–7
torque 253
torque converter 260
torque wrench 51, 54
torsion bar spring 201
torsion bars 207–8
Torx® 52
track rod 216
transaxle 255
transfer box 43
transformer action 167
transmission 89
transmission fluid 258–9
transmission jack 65, 67
transmission systems 86, 89, 247–69
Trapezium 216
trim height 212
triple square screw head 77
trolley jack 66
truss (mushroom) screw head 77
tubeless valve 240
tungsten halogen bulbs 184–5
turbine 260
twin lock nuts 76
twin tube gas damper 210
tyre changer 71, 72
tyre checks, routine 232, 233
tyre inflator 71, 72
tyre markings 241
tyre shop 35
tyre sizes 241–2
tyres 88, 240–3
basic functions 240
requirements 100

Unified Thread Standard 78
universal joints (UJs) 62, 265–6
alignment 265
bearings 265–6
upholstery products 110
used vehicle inspection 98, 99
UV lighting 187

vacuum advance 169
valeting 100–10
data sheets 102–3
exterior cleaning 105–9
interior cleaning 109–10
interior cleaning equipment 104
overview, equipment and safety 102–4
polishing tools 104
pressure washers 103
safety and PPE 102
steam cleaner safety 103–4
valve and brake bands (automatic transmission system) 262
valve clearances 122
valve components 124
valve core 239
valve gear 122–5
valve mechanisms 123
valves 122–5
wheel 239–40
vane air flow meter 150
variable-assistance power steering (VAPS) systems 214
Vehicle and Operator Services Agency (VOSA) regulations 99
vehicle control module (VCM) 143–4
vehicle inspection 95
vernier caliper 61, 62
vinyl, cleaning 109–10
viscosity 138
viscosity index 138
VOSA regulations 99

warm-up time 129
warranty systems 93, 98
washer 78, 79
washing process 105
water jacket 129
water passages 129
water pump 135
water-cooled system 131
wave washer 79
waveform 64
waxes 107
welding process 66
well-base wheel 235
wet vacuum cleaner 104
wheel balancer 71
wheel gun 67
wheel nuts and bolts 236
wheel rim
measurement 238–9
types 239–40
wheel trims 235
wheel valves 239–40
wheels 232, 233, 234
cleaning 105
fitting 238
fixing 236–7
wing covers 97
wing nut 75
wire strippers 56, 57
wires 179
wishbone 201
working environment 18
workshop equipment 65–72
worm and nut steering gear 217, 218
wrenches 51, 76